"This is a thorough investigation of the serious challenges the world faces from health impacts of climate change. Jungmann examines progress made on climate adaptation in different countries, with the benefit of consultation with experts within different levels of government and academia. Individual health risks will vary by country, but global action to adapt to climate change can be greatly facilitated by understanding and building on the variety of knowledge and experience outlined in this type of research."

—Clare Heaviside, Associate Professor, UCL, United Kingdom

"This is a ground-breaking text which examines, for the first time, the interaction between climate change and health adaptation for 192 countries. It examines in detail the drivers that influence the climate change-health adaptation nexus and provides insight into how best society can position itself to cope with the climate emergency."

—John Sweeney, Emeritus Professor, Irish Climate Analysis and Research UnitS (ICARUS), Maynooth University, Ireland

"Climate change already has severe effects on the health of millions of people and its impact will drastically increase in the future. In this timely book, Maximilian Jungmann provides the first global overview of how states perceive health risks associated with climate change and what influences their actions to respond to such risks. His book helps policymakers, businesses, and citizens alike to better understand and prepare for complex, interconnected problems like the climate change and health nexus."

—Bernd Welz, CEO, Klimastiftung für Bürger, Germany

"There is no comparable monograph investigating how states compare in their health adaptation initiatives and which factors influence their perception and respective policies tackling health risks related to climate change. Max Jungmann's work offers an indispensable and comprehensive analysis, using both advanced quantitative and qualitative methods to make sense of the politics of the 'Climate Change-Health Nexus' for scientists, policy makers and the interested public."

—Sebastian Harnisch, Professor for International Relations and Foreign Policy, Heidelberg University, Germany

"This book contributes immensely to the understanding of differential integration of climate change into national public health agendas from a global perspective. Maximilian Jungmann has established the Climate Change and Health Adaptation Index (CHAIn), which may become the tool to evaluate the level of climate resilience of national health systems – a major step towards climate change adaptation for better human health."

—Ina Danquah, Robert Bosch Junior Professor for Sustainable Nutrition in sub-Saharan Africa, Heidelberg University, Germany

The Politics of the Climate Change-Health Nexus

This book compares how governments in 192 countries perceive climate change related health risks and which measures they undertake to protect their populations.

Building on case studies from the United Kingdom, Ireland, South Korea, Japan, and Sri Lanka, *The Politics of the Climate Change-Health Nexus* demonstrates the strong influence of epistemic communities and international organizations on decision making in the field of climate change and health. Jungmann shows that due to the complexity and uncertainty of climate change related health risks, governments depend on the expertise of universities, think tanks, international organizations, and researchers within the public sector to understand, strategize, and implement effective health adaptation measures. Due to their general openness towards new ideas and academic freedom, the book shows that more democratic states tend to demonstrate a higher recognition of the need to protect their populations. However, the level of success largely depends on the strength of their epistemic communities and the involvement of international organizations.

This volume will be of great interest to students and scholars of climate change and public health. It will also be a valuable resource for policymakers from around the world to learn from best practices and thus improve the health adaptation work in their own countries.

Maximilian Jungmann is the co-founder and CEO of Momentum Novum, a sustainability agency in Heidelberg and Quito, and a researcher at Heidelberg University. He holds a PhD in Political Science from Heidelberg University and does research on the politics of climate change and sustainable development.

Routledge Studies in Environment and Health

The study of the impact of environmental change on human health has rapidly gained momentum in recent years, and an increasing number of scholars are now turning their attention to this issue. Reflecting the development of this emerging body of work, the *Routledge Studies in Environment and Health* series is dedicated to supporting this growing area with cutting edge interdisciplinary research targeted at a global audience. The books in this series cover key issues such as climate change, urbanisation, waste management, water quality, environmental degradation and pollution, and examine the ways in which these factors impact human health from a social, economic and political perspective.

Comprising edited collections, co-authored volumes and single author monographs, this innovative series provides an invaluable resource for advanced undergraduate and postgraduate students, scholars, policy makers and practitioners with an interest in this new and important field of study.

Healthy Urban Environments
More-than-Human Theories
Cecily Maller

Climate Information for Public Health Action
Edited by Madeleine C. Thomson and Simon J. Mason

Environmental Health Risks
Ethical Aspects
Edited by Friedo Zölzer and Gaston Meskens

Climate Change and Urban Health
The Case of Hong Kong as a Subtropical City
Emily Ying Yang Chan

Environmental Health and the U.S. Federal System
Sustainably Managing Health Hazards
Michael R. Greenberg and Dona Schneider

The Politics of the Climate Change-Health Nexus
Maximilian Jungmann

For more information about this series, please visit: https://www.routledge.com/Routledge-Studies-in-Environment-and-Health/book-series/RSEH.

The Politics of the Climate Change-Health Nexus

Maximilian Jungmann

Routledge
Taylor & Francis Group

LONDON AND NEW YORK

First published 2021
by Routledge
2 Park Square, Milton Park, Abingdon, Oxon OX14 4RN

and by Routledge
52 Vanderbilt Avenue, New York, NY 10017

Routledge is an imprint of the Taylor & Francis Group, an informa business

British Library Cataloguing in Publication Data
A catalogue record for this book is available from the British Library

Library of Congress Cataloging-in-Publication Data
Names: Jungmann, Maximilian, author.
Title: The politics of the climate change-health nexus / by Maximilian
　Jungmann.
Description: Abingdon, Oxon ; New York, NY : Routledge, 2021. |
　Series: Routledge studies in environment and health | Includes
　bibliographical references and index.
Identifiers: LCCN 2020051452 (print) | LCCN 2020051453 (ebook) |
　ISBN 9780367703134 (hardback) | ISBN 9781003145646 (ebook)
Subjects: LCSH: Public health–Environmental aspects. | Climatic
　changes–Health aspects. | Health risk assessment. | Medical
　climatology–Political aspects.
Classification: LCC RA793 .J84 2021 (print) | LCC RA793 (ebook) |
　DDC 362.1–dc23
LC record available at https://lccn.loc.gov/2020051452
LC ebook record available at https://lccn.loc.gov/2020051453

ISBN: 978-0-367-70313-4 (hbk)
ISBN: 978-0-367-70314-1 (pbk)
ISBN: 978-1-003-14564-6 (ebk)

Typeset in Goudy
by Taylor & Francis Books

In memory of Kurt Jungmann

Contents

x *Contents*

Figures

Tables

List of abbreviations

CCC	Committee on Climate Change United Kingdom
CHAIn	Climate Change and Health Adaptation Index
COP	Conference of the Parties
CPI	Corruption Perception Index
DCCAE	Department of Communications, Climate Action and Environment Ireland
EU	European Union
GDP	Gross Domestic Product
KEI	Korea Environmental Institute
IPCC	Intergovernmental Panel on Climate Change
LNA	Large-N-Analysis
LSHTM	London School of Hygiene and Tropical Medicine
MMR	Mixed Methods Research
NCs	National Communications
ND-GAIN	Notre Dame Global Adaptation Initiative
NGOs	Non-Governmental Organizations
OECD	Organisation for Economic Co-operation and Development
PPP	Purchasing Power Parity
RoK	Republic of Korea
SNA	Small-N-Analysis
UK	United Kingdom
UKCP	UK Climate Projections
UN	United Nations
UNDP	United Nations Development Programme
UNFCCC	United Nations Framework Convention on Climate Change
WHO	World Health Organization
WMO	World Meteorological Organization

Acknowledgements

PhDs are often described as long journeys with many highs and lows. This PhD, however, felt more like an academic triathlon. It was a continuous challenge to develop and adhere to an innovative and coherent research design that a) integrates different academic disciplines, b) collects data and seeks to contribute to a better understanding of a constantly changing and rapidly developing, highly complex and uncertain research field, and c) mixes various research traditions and methods to benefit from their advantages and to balance out their individual weaknesses. This triathlon has helped me to grow as a person by learning from some of the world's greatest minds and it has brought me to five different countries where I had the opportunity to gain insights into various perspectives on the same research subject. Although this academic triathlon was one of the greatest challenges I have ever encountered, it simultaneously was one of the most rewarding projects I have had the chance to work on.

This unique opportunity would not have been possible without the constant support, creativity, humor, academic brilliance, and empathy of my supervisor, Prof. Dr. Sebastian Harnisch. Through his trust in my abilities, his openness towards new academic pathways, and especially his outstanding support throughout this entire project, I would not have had the power, motivation, or mental capacity to make it to the finish line. Prof. Harnisch has had a great impact on the conceptual framework of this thesis, the theoretical and methodological approach, and the overall quality of the entire research project. Without him, this thesis would not exist in this form.

During my research trip to Ireland I had the great opportunity to meet with Prof. John Sweeney, PhD, Ireland's most renowned climate scientist, expert on climate change and health, and outstanding educator and mentor. Not only did Prof. Sweeney share his knowledge and expertise on climate change and health with me, but he also volunteered to be the external supervisor of this thesis. Prof. Sweeney's tremendous support and encouragement have helped me to better understand the drivers of and barriers to health adaptation to climate change in various countries. His kindness, intelligence, and constructive feedback have motivated me to continue with this academic triathlon, especially at times when the finish line seemed far away.

Furthermore, Prof. Dr. Jale Tosun, Prof. Dr. Manfred G. Schmidt, Dr. Sanam Vardag, Dr. Clare Heaviside, Kusum Athukorala, Prof. Dr. Yasushi Honda, Dr. Ina Kelly, Dr. Jordan Kassalow, and Dr. Johannes Gabriel had a significant impact on the success of this project by providing me with their valuable insights and feedback. They have encouraged me to discover new academic paths and assured me of the relevance of this study.

I would further like to express my sincere gratitude towards the German Federal Foundation for the Environment (DBU) and especially my supervisors at the foundation, Verena Exner and Dr. Hedda Schlegel-Starmann. Without the financial and ideational support of the DBU I would not have had the necessary resources to finish this research project. Over the course of the last three years, the DBU's interdisciplinary network and strong research focus have provided me with the necessary skillset and understanding of interdisciplinary research that was pivotal for the success of this study. Moreover, I am very grateful for the support from my colleagues, friends, and mentors at the Heidelberg Center for the Environment (HCE), especially Prof. Dr. André Butz, Prof. Timo Goeschl, PhD, Prof. Dr. Werner Aeschbach, and Dr. Nicole Aeschbach, and our amazing team at Momentum Novum.

I would further like to thank my dear friends Jakob Landwehr, Michael Valdivieso Muñoz, Chris McKenna, Tsesa Monaghan, Asra Shakoor, Max Lacey, Andrea Wong, Saeko Yoshimatsu, Lauren Kiser, Eileen Austin, and Angelina Pienczykowski for their outstanding support along the way. They have provided exceptionally valuable feedback on the distinct chapters of this thesis and constantly encouraged me to continue with my work.

Moreover, I am eternally grateful for the support, love, and patience of my parents. The last three years have not always been easy and this PhD has taken a large toll on the time I could have spent with them. I very much appreciate the unlimited support of my parents, grandparents, and my entire family, especially my cousin Philipp. It is almost impossible to fully understand the scale and intensity of a PhD if you are not working on it yourself, but they always showed great interest in my work and supported me along the way.

Most importantly, I would like to thank my partner, Martina. There are no words to adequately express my gratitude towards her since she is the most important reason for the existence of this book. She always goes absolutely above and beyond everything to support me. From lengthy but entertaining discussions about the conceptual, methodological, theoretical, and empirical framework of this thesis, to proof-reading this entire book, there is nothing Martina did not do to help me get over the finish line. Most importantly, she encouraged me to keep going in difficult situations, showed compassion at times of frustration, and helped me to find solutions for any issue that came up. Martina is the greatest inspiration and role model one can wish for. Thank you for everything.

Preface

At the time when I finally submitted this book to Routledge, the world had become a completely different one than when I was doing the vast part of my research. With COVID-19, a global pandemic is now dominating the world. People are socially distancing from each other, borders are closed, economic crises have erupted, and the political agenda and media coverage have shifted to crisis mode. At the same time, the topic of climate change and health has become more important than ever as we now see, hear, and feel the massive consequences of health risks. Compared with COVID-19, however, climate change related health risks represent an even broader and more complex challenge to decision makers as most health risks that can be associated with climate change are more diffuse, interconnected, and difficult to respond to. Moreover, in addition to some very direct primary health risks, many climate change effects constitute slow-onset events and thus require completely different capacities to adapt than the crisis response the world is currently setting in place as a consequence of COVID-19.

Much more research is needed to understand current future risks and what can and needs to be done to protect human and planetary health. This book seeks to shed light on a largely under-researched area that, however, has a strong impact on all other research areas: the politics of the climate change and health nexus. To date, not much is known about how different states across the globe perceive climate change related health risks, how they adapt to such risks and what drives their actions. Therefore, this book seeks to provide a global overview of national-level health adaptation measures through the introduction of the Climate Change and Health Adaptation Index (CHAIn) and a better understanding of the drivers and barriers of health adaptation through the Sieve Model on Climate Change and Health. Due to the novelty and rapid development of the research subject, this study is designed to provide a starting ground for further research projects that take into account different governance levels, such as the regional and local level, and longer time periods through longitudinal analyses. Ultimately, it is the goal of this book to foster better informed and more comprehensive adaptation and mitigation actions and to contribute to a physical and social environment that is sustainable and rests on international cooperation.

Part I
Introduction

1 Between crisis and long-term challenge

"The Climate Crisis is a Public Health Crisis"

– Al Gore[1]

In 2017, a major dengue outbreak hit Sri Lanka (WHO 2017). Between January and July 2017, the Sri Lankan government reported more than 80,000 infections, including 215 deaths (WHO 2017). In summer 2018, almost 80 people died and more than 30,000 had to be taken to hospital due to heat exhaustion and heat stroke when a massive heatwave hit Japan (Sim 2018). The same heatwave led to more than 40 deaths in South Korea and severe crop destruction in North Korea (Haas 2018). Similarly, people in Australia had to cope with severe heat, droughts, and forest fires in 2018 and the beginning of 2019, with temperatures of over 50 °C (Cox and Watts 2019). At around the same time, the polar vortex led to temperatures of down to −18 °C in the United States of America and Canada, which resulted in more than 20 deaths and hundreds of cases of frostbite (BBC 2019b, Rowlatt 2019, Milman 2019). In April 2019, two massive storms, cyclones Idai and Kenneth, hit Mozambique and Zimbabwe, leading to thousands of deaths and injuries, destroying great parts of the infrastructure, and causing severe cholera and malaria outbreaks (Taube 2019). At the time of the submission of this thesis, by the end of June 2019, a major heatwave arrived in Europe and led to temperatures of up to 45.9 °C in France (BBC 2019a). When revising this book, at the end of 2019 and the beginning of 2020, Australia experienced an exceptionally high number of devastating bush fires that destroyed more than 11 million hectares of land and pushed numerous endangered species to extinction (Morton 2020, BBC 2020).

These examples constitute only a small and random sample of the major extreme weather events in recent years. Whilst they are from different regions across the globe and have led to different health risks, they all have one aspect in common: climate change is very likely to be one driver behind them. The examples show that climate change is no longer only an abstract risk in the future, but its consequences are real and can be felt already, especially when it comes to climate change related health risks. At the same time, however, the nexus between climate change and health is a very complex one and contains high levels of uncertainty. Although the climate has already started to change, the major effects will affect humanity in the future, and we do not know the exact scale and intensity of future climate change related health risks. To provide an overview of what we do know already, the following section summarizes the current state of research on climate change and health.

A complex relationship

Although research on climate change and health has rapidly increased over the last years, many unknowns persist, in particular when it comes to understanding and measuring the concrete impact of climate change on certain health risks. In addition to climate change, health risks often depend on a great number of other complex global factors that are not directly related to climate change, such as globalization or migration, which makes it difficult to state that climate change is the main reason for certain health risks. What researchers from all over the world have found, however, is that climate change does have a significant effect on a great variety of health risks (Kovats et al. 2003, McMichael 2014, Patz et al. 2014). This chapter seeks to provide an overview of what academia does and does not know about the nexus between climate change and health and why adapting to climate change related health risks is a complex challenge that is often called a "wicked problem" (Fermeer, Dewolf, and Breeman 2013, 27).[2]

Assessing the challenge

Although in many cases it is not possible to clearly state that climate change directly leads to certain health risks, the vast majority of researchers on the subject agrees that anthropogenic climate change has a severe impact on health and specific health determinants, such as access to clean water and fair sanitation.[3] The health risks associated with climate change, however, also depend on a number of other factors, such as globalization, migration, and economic development, that may again influence each other (Neira et al. 2008, 424, Patz et al. 2012, 3). At the same time, it is clear that climate change increases the possibility of certain negative health effects. They range from very direct consequences of climate change, such as risks caused by floods and storms, to very diffuse processes, which in the long run alter pivotal determinants of health (McMichael 2013).

As Figure 1.1 illustrates, the relationship between climate change and health is constituted by an interdependent web of different factors that either directly or indirectly influence each other. The great amount and complexity of health risks that can be associated with climate change can be overwhelming at first sight. Therefore, this chapter introduces the trichotomy of primary, secondary, and tertiary climate change related health risks,

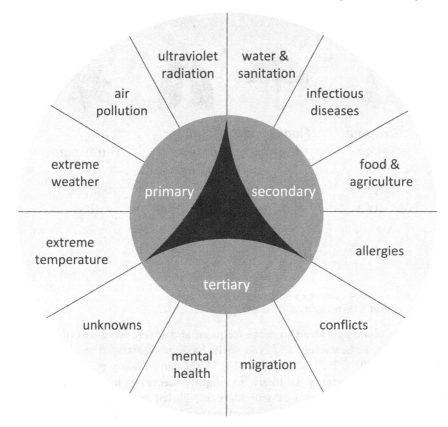

Figure 1.1 Impact of climate change on human health
Source: author's own figure.

a categorization established by McMichael (2013, 1338). While other authors simply distinguish between short-term and long-term health risks, the trichotomy allows a much deeper understanding of the health risks associated with climate change and makes it easier to identify the respective causes for each risk (McMichael 2013, WHO/UN 2009, WHO/WMO 2012).

Primary risks

Primary climate change related health risks are direct consequences of climate change (Gislason 2015, 540, McMichael 2013, 1338–1339) They often have an immediate impact on people's health and can be very dangerous, especially for elderly people (ibid). They span from extreme temperature (heat and cold), extreme weather events, such as floods, storms, droughts, and fires, to exacerbated effects of indoor and outdoor air pollution, and ultraviolet (UV) radiation (Figure 1.2).

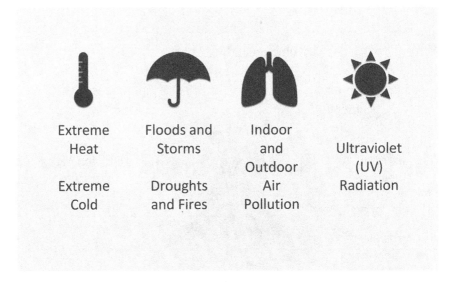

Figure 1.2 Primary climate change related health risks
Source: author's own figure.

Climate change can lead to more frequent and severe instances of extreme temperature, as heatwaves and cold spells have demonstrated in recent years (Casas et al. 2016, 75, Hajat et al. 2014). Although in many places across the world winter mortality is likely to slightly decrease in years to come, because winter temperatures will increase, global warming may not only trigger heatwaves, but also periods of extremely cold weather, as the polar vortex showed in Northern America at the beginning of 2019 (Milman 2019, Gasparrini et al. 2017). The direct effects of climate change, especially extreme heat, affect tropical regions the most (Schleussner et al. 2016, Vicedo-Cabrera et al. 2018).

Despite regional differences, most parts of the world will experience some effects of climate change. Mora et al. (2017, 501) found:

> Around 30% of the world's population is currently exposed to climatic conditions exceeding this deadly threshold for at least 20 days a year. By 2100, this percentage is projected to increase to ~48% under a scenario with drastic reductions of greenhouse gas emissions and ~74% under a scenario of growing emissions.

Whether the international community will achieve the 1.5 °C goal, which was set out in the Paris Agreement, will massively affect heat-related mortality (Mitchell et al. 2018). Mitchell et al. (2018) project that "in key European cities, stabilizing climate warming at 1.5 °C would decrease extreme heat-related mortality by 15–22% per summer compared with stabilization at 2 °C."

Heatwaves and extremely cold weather both can have severe effects on humans' well-being (Casas et al. 2016, 74–75, Leal Filho, Azeiteiro, and Alves 2016b, 5). Extreme heat can lead to heat cramps, heat edema, heat exhaustion (which may be the early stage of a heat stroke), heat stroke, and dehydration (Margolis 2013, 97, Rom and Pinkerton 2013, 10).[4] Especially in cities, where the "urban heat island effect"[5] can lead to more than 5 °C higher temperatures compared with rural areas, extreme heat is and will be a major health risk, especially for elderly people, homeless people, and pregnant women (WHO/WMO 2012, 40, Heaviside, Macintyre, and Vardoulakis 2017, Leal Filho, Azeiteiro, and Alves 2016a, 5). The World Health Organization and World Meteorological Organization (WHO/WMO 2012, 40) estimate that "by the 2050s, heat events that would currently occur only once every 20 years will be experienced on average every 2–5 years." Similar to severe health effects due to heat, extremely cold weather can imply a great number of serious health risks, such as frostbite or death from exposure (Anderson and Bell 2009, Leal Filho et al. 2016b, 5).[6] Extreme heat or cold can further trigger or lead to cardiovascular diseases and reinforce already prevalent diseases such as respiratory disorders (Casas et al. 2016, 74–75, Leal Filho et al. 2016b, 5). According to Friel et al. (2011, 134–135), climate change may result in a higher risk of cardiovascular diseases through three pathways: directly via extreme temperatures and air pollution, and indirectly via "changes to dietary options." High temperatures can lead to "increased core body temperature, increased heart rate, shift of blood flow from central organs to skin, increased sweating, and associated dehydration if sufficient replacement liquid is not consumed" (Friel et al. 2011, 135). Climate change may affect air pollution due to its effects on weather and pollution concentrations, human-induced emissions, and the distribution of airborne allergens (Friel et al. 2011, 135). Moreover, extreme heat often leads to decreased mobility and higher consumption of processed foods, which can contribute to cardiovascular diseases (Friel et al. 2011, 135).

In addition to short-term impacts of warm weather, climate change can also lead to droughts and forest fires, which can again cause numerous health risks, such as thermal injuries or more diffuse health risks, as for example detrimental effects on mental health (Morita and Kinney 2014, 114, Shukla 2016, 16). Already in 2007, Working Group II to the Fourth Assessment Report of the Intergovernmental Panel on Climate Change (IPCC) stated that climate change is likely to result in an increased number of deaths due to heatwaves, fires, and other health risks (Confalonieri et al. 2007, 393) Past events, such as the great heatwave in Europe in 2003, in which more than 70,000 people died, and constantly increasing temperatures across a great number of countries, have demonstrated how extreme weather events threaten human health (WHO/WMO 2012, 40). According to recent studies, especially women, children, elderly people, indigenous people, and those living with fewer financial resources belong to vulnerable groups that are at heightened risk during heatwaves or other extreme weather events (Casas et al. 2016, 73–74, Setti et al. 2016).[7]

Recent studies show that climate change contributes to an increase in the number and intensity of extreme weather events across the globe (Balbus et al. 2008, Cann et al. 2013, Stott 2016). Heavy rainfall, storms, coastal flooding in combination with sea-level rise, and other weather extremes can pose severe health risks to humans (Luber et al. 2014, 221, Lloyd et al. 2016). Heavy rainfall and floods can lead to increasing numbers of people who die from drowning or hypothermia if they are not rescued in time (WHO/WMO 2012, 26). Storms can directly lead to injuries or deaths if trees or buildings collapse or debris whirls through the air (Shumake-Guillemot, Villalobos-Prats, and Campbell-Lendrum 2015, 9, Watts et al. 2015, 4). Watts et al. (2015, 4) state that the number and intensity of floods and storms is likely to increase in the future due to climate change. The World Health Organization (WHO) expects that, as floods increase, the number of deaths from exposure or drowning will likely increase as well (WHO/WMO 2012, 26).[8] Especially women and people at younger ages are at risk from extreme weather events, such as storms, hurricanes or extreme precipitation (WHO n.d.-a, 3).

Access to clean air, an essential determinant of health, is also at risk as climate change accelerates (Ojeh and Aworinde 2016, 173). Climate change contributes to higher temperatures in various regions across the globe, which is likely to result in increasing levels of urban air pollutants since they are sensitive to rising temperatures (Gislason 2015, 540). Rising temperatures directly impact the "levels of pollutants that are formed and the ways these pollutants are dispersed" (Patz et al. 2012, 23). Especially ozone levels often increase with rising temperatures, as was the case during heatwaves in various regions across the world in recent years (Patz et al. 2012, 23). According to the World Health Organization, ozone-related deaths from climate change may increase by 4.5 percent by the mid 2050s compared with levels from the 1990s (Patz et al. 2012, 23). In addition to negative effects on outside air quality, climate change may also affect humans' exposure to air pollutants inside houses since extreme temperatures and extreme weather events may lead to people spending more time indoors where they are more frequently exposed to house dust mites, mold spores, allergens, and material-based air pollutants such as fumes from paint or linoleum or products of combustion from heating fuels used for warming and cooling (Beggs 2016, 1, Ziska 2016, 130). As a consequence, asthma prevalence and other respiratory diseases are expected to increase with global warming (Luber et al. 2014, 222).

Additionally, climate change directly influences people's exposure to ultraviolet (UV) radiation since increasing temperatures in many countries have led people to spend more time outside (Smith et al. 2014, 722).[9] Outdoor exposure is mainly increased for people who have limited choice but to be outdoors and possess limited protection against UV. Increased exposure to UV radiation may lead to a higher prevalence of skin or eye diseases (Smith et al. 2014, 722). According to Smith et al. (2014, 722) increased levels of UV radiation "and maximum summertime day temperatures are related to the prevalence of non-melanoma skin cancers and cataracts in the eye." According to Bharath and Turner (2009), 90 percent of skin cancers are non-melanocytic. Additionally, evidence from Ireland suggests that with warmer summers people tend to barbecue more and consume more alcohol, which involves additional health risks such as respiratory diseases due to increased exposure to fumes, increased rates of colorectal cancer from processed meats, as well as behavioral risks as a consequence of high alcohol consumption (Technical University of Dublin, personal communication, 24 November 2018).

Secondary risks

Secondary risks influence essential determinants of health (McMichael 2013, 1338). They include biological, physical, and ecological alteration processes, such as the impact of climate change on food production, access to potable water, and the intensity and range of infectious diseases (McMichael 2013, 1338) (Figure 1.3). These risks can arise from extreme weather events or from long-term processes in which health is altered due to changing climatic conditions (McMichael 2013, 1338).

Water and sanitation are fundamental pillars of health. Access to clean water is the key condition for life on earth since most organisms simply do not function without a sufficient amount of water (WHO n.d.-b). Climate change threatens access to clean drinking water and sanitation in numerous countries across the globe (Shukla 2016, 16). Changing precipitation patterns

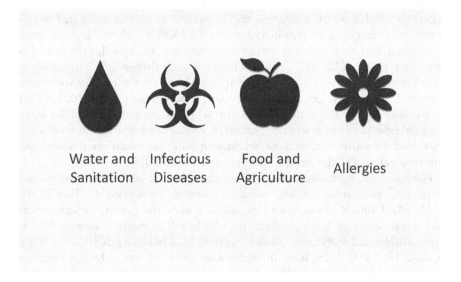

Water and   Infectious   Food and
Sanitation   Diseases   Agriculture   Allergies

Figure 1.3 Secondary climate change related health risks
Source: author's own figure.

and reduced water availability may lead to water-borne diseases, such as harmful algal blooms, cholera, dysentery, typhoid, diarrheal diseases, and leptospirosis (Nichols et al. 2018, 1, Aparicio-Effen et al. 2016, 231, Vezzulli, Colwell, and Pruzzo 2013, 817). Similarly, extreme weather events can affect access to sanitation, and inadequate sanitation can in turn contribute to the spread of numerous diseases, including cholera and diarrheal diseases (WHO/WMO 2012, 12). Massive flooding in India, China, Brazil, and other countries, as well as tropical storms and hurricanes, have demonstrated how extreme weather can jeopardize access to potable water (Shukla 2016, 16). Furthermore, flooding can lead to the contamination of drinking water since runoff from sewage lines or pollution can more easily flush into those waters (Patz 2013, 230). Several regions across the globe are already struggling with the impact of climate change on water-borne transmittable diseases, diarrheal diseases, and similar health risks (Aparicio-Effen et al. 2016, 254, Patz 2013, 230). According to Patz (2013, 221), heavy rainfall can also lead to public health risks in recreational waters, for instance near beaches in coastal areas, since extensive runoff can cause higher bacterial levels. Additionally, rising sea levels and ocean acidification, both of which are strongly influenced by climate change, pose existential threats to almost all countries across the globe, but most of all small island development states (Akpinar-Elci and Sealy 2013, 281). In addition to directly threatening coastal livelihoods, rising sea levels and ocean acidification can result in the salination and contamination of coastal freshwater aquifers and disrupt water treatment services (Patz et al. 2012, 3).

In addition to excessive amounts of non-potable water, water scarcity can result in severe health risks. Droughts frequently result in health issues since access to potable water, sanitation and hygiene is more restricted and people resort to consuming contaminated water (WHO/UN 2009, 2). Such practices often lead to numerous waterborne diseases, such as diarrhea or cholera (Patz et al. 2012, 4).[10] The WHO estimates that in 2030, the risk of diarrhea will in some regions be 10 percent higher than if no climate change occurred (Campbell-Lendrum, Corvalán, and Prüss-Ustün 2003, 154). Moreover, due to the sensitivity of the schistosome parasite and its intermediate host snails to water temperatures, higher temperatures in freshwater may lead to more instances of schistosomiasis, especially in tropical and sub-tropical areas (McCreesh, Nikulin, and Booth 2015).

Furthermore, agriculture highly depends on water resources, arable land, fertilizers, and, most of all, suitable climatic conditions (WHO/WMO 2012, 34). Climate change can dramatically alter the quality of agriculture and the quantity of food production, which will in many countries lead to malnutrition and starvation (Butler 2014b, 130, Chabejong 2016, 133, Watts et al. 2015, 1863). Long-term limitations in access to water, higher temperatures, as well as extreme weather events such as storms, extreme precipitation, floods, or droughts massively impact the agricultural sector (Casas et al. 2016, 74, Hutton 2011, 1, Patz et al. 2014, 1565). As a consequence, there may be a lack of food supply and food prices can rapidly rise, making it impossible for people to get sufficient food (Singh and Rao 2014, 168). There is therefore an increased chance of malnutrition, poor nutrition, or other health problems stemming from caloric deficiencies due to climate change, leading to further strain on global health systems (Confalonieri et al. 2007, 393, WHO/WMO 2012, 30).[11] People in Sub-Sahara Africa and South-East Asia and poor people in rural communities in general will suffer the most as climate change continues to affect the quality and quantity of food production (Fanzo et al. 2018). Extreme heat events and droughts, such as the one Europe experienced in 2003, will increase fivefold and possibly even tenfold until the year 2050, leading to bad harvests and starvation crises (Machalaba 2015, 228, Hajat et al. 2014).

Furthermore, the spread of numerous infectious diseases is directly affected by climatic factors (Leal Filho et al. 2016b, 6, Ogden 2014).[12] According to Veenema et al. (2017, 629), dengue, yellow fever, west Nile virus, Japanese encephalitis, ross river virus, and malaria are among the vector borne diseases that spread more rapidly as a consequence of climate change. Moreover, due to climate change, vector borne diseases increasingly spread to regions that have previously been unaffected (McMichael 2013, 1340, Semenza 2014, 7347. A key factor that influences the spread of vector borne diseases is temperature (Ojeh and Aworinde 2016, 174). With rising temperatures, many formerly unaffected areas will have to deal with vector borne diseases as reproductive patterns are changing due to climatic variation and the novel geographic spread of vectors (Rom and Pinkerton 2013, 12).[13] For the dengue fever for example, which is transmitted primarily through the Aedes Aegypti mosquito, scientists expect high increases in the geographic spread of the disease as a consequence of climate change (Patz et al. 2012, 4). Rom and Pinkerton (2013, 12) state that with a temperature increase of 5 °C, the number of eggs laid per female mosquito is doubled.[14] Additionally, the number of eggs laid per mosquito greatly increases with humidity levels of more than 60 percent (Rom and Pinkerton 2013, 12). Moreover, the incubation time for eggs inside mosquitoes reduces with higher temperatures, which ultimately leads to the reproduction of more mosquitoes in less time (Gillis 2016, Rom and Pinkerton 2013, 12). Last but not least, most mosquitoes use standing waters or stagnant areas of moving water to hatch their eggs. Therefore, increased rainfall can help mosquitoes to alter their life cycles by having more standing waters available (Rom and Pinkerton 2013, 12). Since climate change involves rising temperatures, more humidity, and more extreme weather events, including flooding, it has a strong influence on the geographic spread of mosquitoes (Rom and Pinkerton 2013, 12). The WHO has found out that the number of dengue transmitting mosquitoes has increased 30-fold over recent decades with now 50 to 100 million cases around the world (Cromar and Cromar 2013, 167). While dengue was once only prevalent in some tropical areas, it can nowadays be found in more than 100 countries (Cromar and Cromar 2013, 167).

Additionally, Asad and Carpenter (2018, 38) found out that climate change affects some of the factors that contribute to the spread of the zika virus, such as temperature and precipitation. As climate change further advances, the number of areas that are affected by vector borne diseases is expected to grow (Pereda and Alves 2016, 326). In combination with globalization and other factors, climate change therefore leads to an increased number of infectious diseases across the globe and possibly new epidemics and pandemics (Patz et al. 2003, 112, WHO 2003, 11).

Moreover, climate change influences the genetic predisposition towards certain hypersensitivity reactions: allergies (Beggs 2014, 107, Ziska and Beggs 2012, 31). While research on this subject is still developing, it is already clear that climate change does affect plant biology and thus increases the risk of allergic reactions (Beggs 2014, 111). According to Beggs (2014, 106), changing temperatures and precipitation patterns are likely to impact "pollen production and atmospheric pollen concentration, pollen season, plant and pollen spatial distribution and pollen allergenicity."

Allergies can have many forms, yet among the most prevalent ones are reactions to pollen in air, which can also exacerbate asthma (WHO/WMO 2012, 48).[15] The World Health Organization and World Meteorological Organization estimate that globally 235 million people suffer from asthma and about 80 million adults in Europe (over 24 percent of the adults living in Europe) are affected by allergies (WHO/WMO 2012, 48). The numbers of globally affected children are even higher, with 30–40 percent of them suffering from allergies (WHO/WMO 2012, 48). Climate change is not the only factor that leads to an increase in allergic diseases, but it does have a significant impact and is connected to many other factors (Confalonieri et al. 2007, 393). According to the hygiene hypothesis, for example, more and more people have allergies because they are exposed to fewer allergens as children (Okada et al. 2010). Climate change can also lead to more time spent indoors due to extreme temperatures and extreme weather events, which may again lead to decreased exposure to allergens as a child. Regarding the future of these diseases, it is estimated that hay fever, asthma, and reactions to ragweed and other pollens are very likely to increase, leading not only to human suffering but also to financial losses (Rom and Pinkerton 2013, 11). Lake et al. (2017, 385) estimate that "sensitization to ragweed will more than double in Europe, from 33 to 77 million people, by 2041–2060," with major increases in countries where sensitization is currently uncommon, such as Germany or Poland.

Tertiary risks

Tertiary risks are very diffuse effects of climate change, which for some observers may on first sight be difficult to identify (Gislason 2015, 540, McMichael 2013, 1338). They include, but are not limited to, health risks caused by conflicts, migration, displacement, and mental health disorders due to various climate change related effects (Gislason 2015, 540, McMichael 2013, 1338, McMichael, Barnett, and McMichael 2012). Tertiary climate change related health risks are often linked to a great number of other health risks and it is not always possible to clearly state which risks lead to which effect. However, the general impact of climate change on these risks cannot be denied.

Climate change has the potential to spark or intensify national, regional, or international conflicts (Bowles, Braidwood, and Butler 2014, 151, Butler 2014b, 132, Mach et al. 2019). The potential reasons for such conflicts are as plentiful as diverse. Among the most straightforward connections between climate change and conflicts is the impact of a changing climate on resource availability (Bowles, Braidwood, and Butler 2014, 147). On the one hand, increasing temperatures may lead to melting ice and glaciers which can facilitate access to non-renewable resources, such as natural gas, crude oil or minerals (Bowles, Braidwood, and Butler 2014, 147). On the other hand, vital resources, such as potable water, fertile soil, or fishing grounds are likely to deplete in various regions across the globe as a consequence of droughts and increasing temperatures or constantly shifting precipitation patterns (Butler 2014a, 3). Both instances may foster conflicts, yet in different ways. In the first case, previously unpopulated international territories or resources, as for instance in the Arctic, are at risk of becoming contested (Lamothe 2018). In the second case, droughts, floods, and other climate change related agricultural challenges can result in famines and clashes over remaining resources (Butler 2014b, 132, Patz et al. 2012, 3).

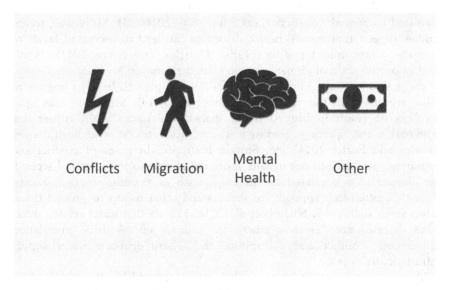

Figure 1.4 Tertiary climate change related health risks
Source: author's own figure.

Conflicts pose severe threats to human health and existence, no matter whether they are within or between states. A study from 2009 estimates that more than 2.7 billion people in over 46 countries will experience a very high risk of violent conflict resulting from climate change in combination with social, economic, and political problems (Barbara 2013, 89). In addition to very direct health risks originating during armed conflicts, such as deaths and injuries, public health services are often limited or unavailable during conflicts and access to essential resources and health services is also limited (Barbara 2013, 87, Watts et al. 2015, 4). Furthermore, violent conflicts often result in migration and an increase in mental health disorders.

Migration is among the most frequent responses to conflicts. When conflicts between or inside states become violent, great parts of the population are forced to leave their homes and migrate to safer places (Barbara 2013, 81, Patz 2013, 219, Abel et al. 2019). Additionally, severe and constant flooding, long-term droughts, depleting resources, such as water or timber, or other climate change related consequences for human habitats may result in migration movements (Grover 2013, 14). Moreover, as climate change progresses, rising sea levels, the salination of groundwater resources, the destruction of farmland or housing areas, and decreasing fishing grounds force more and more people to migrate, which poses a number of health risks (Bowles, Reuveny, and Butler 2014, 137, Leal Filho et al. 2016b, 3).

Forced migration often involves an increasing prevalence of infectious diseases due to lacking hygienic conditions or access to potable water and sanitation. At the same time, it can cause stress and anxiety, which can in turn lead to mental disorders (Leal Filho et al. 2016b, 3). Moreover, recent studies suggest that mental health disorders can lead to increased levels of immune compromised people (Blume, Douglas, and Evans 2011). While studies on the scale of climate change related migration are contested, some of them assume that 250 million to 1 billion people will have to migrate to other regions until the year 2050 (Grover 2013, 14). Migration may spur conflicts or result in limited living spaces, which can again trigger discrimination and uprisings, leading to mental disorders or other health issues (Bowles and Butler 2014, 586, Shukla 2016, 19). In times of conflict and migration, populations are usually more vulnerable to primary and secondary climate change related health risks, such as extreme weather events, since they often lack appropriate shelters and other means to protect themselves from such risks. Nichols et al. (2018, 1) state that water-related infectious diseases are "greatest where the climate effects drive population movements, conflict and disruption, and where drinking water supply infrastructure is poor."

Migration and conflicts may result in severe mental health risks, such as post-traumatic stress disorder (PTSD) (Smith et al. 2014, 720, Veenema et al. 2017, 629). Moreover, recent studies have shown that extreme weather events, such as extreme heat or floods, often result in trauma, chronic stress, anxiety, and depressions, especially in communities where people lost properties or friends and family members due to extreme weather events (Barbara 2013, 83, Doherty and Clayton 2011, Shukla 2016, 17). According to Shukla (2016, 17), individuals who already have mental disorders are particularly at risk of suffering heat related injuries. Furthermore, an increasing number of studies claims that suicide rates may grow since the effects of climate change may lead to increased prevalence of depression and mental illness (Luber et al. 2014, 228). Increasing temperatures, changing weather patterns, and extreme weather events may directly lead to higher suicide rates or they may cause destruction of housing or farmland and ultimately lead to suicides (Berry, Bowen, and Kjellstrom 2008, 123, Smith et al. 2014, 720). There is evidence that, in both developed and developing states, droughts have increased suicide rates of male farmers (WHO n.d.-a, 3).

In addition to influencing conflicts, migration, and mental health risks, climate change has the potential of triggering several other factors that may negatively impact pivotal determinants of health. Such risks include economic losses due to climate change related extreme weather events or extreme temperatures, including people losing their homes or jobs and thus suffering from economic hardship (WHO/UN 2009, 2, Smith et al. 2014, 713). This again may result in worsened health conditions since people may no longer be able to afford sufficient nutrition or health services. Additionally, climate change related extreme weather events may cause enormous destruction in public infrastructure or social services, which can lead to lacking access to both primary and emergency care services (Barbara 2013, 80).

The bigger picture

Climate change affects health in multiple direct or indirect ways. While research on the subject is still growing and not all effects of climate change on health have already been discovered, it is very certain that all countries need to adapt to climate change related health risks since global mitigation efforts have not been successful enough yet and the effects of climate change on health are already noticeable across the globe. With regards to current greenhouse gas emission trends, climate change will rapidly gain speed and massively affect the health of all populations. Although the specific health risks associated with climate change differ from region to region, all parts of the world will be affected – in one way or the other. Accordingly, adaptation becomes inevitable.

But what can states do to manage these risks? What does effective adaptation to climate change related health risks look like and what are necessary actors and factors for successful adaptation?

Notes

1 Al Gore made this statement before his one-hour special of 24 hours of reality, Al Gore discussed climate change related health risks: https://twitter.com/algore/status/1075797921608712193?lang=en.
2 For further information on the term "wicked problem" see page 24.
3 See among others: (Charron et al. 2008, Gislason 2015, Patz 2018, Patz and Thomson 2018, Hales et al. 2014, Hales et al. 2002, Patz et al. 2012, Singh et al. 2001, Hales, Edwards, and Kovats 2003, McIver et al. 2016, Bambrick and Hales 2014, McMichael, Woodruff, and Hales 2006, Schnitter et al. 2018, Hess, Schramm, and Luber 2014, Hess et al. 2013, Luber et al. 2014, Hess, Malilay, and Parkinson 2008, Louis and Hess 2008, Luber et al. 2014, Vicedo-Cabrera et al. 2018, Gasparrini et al. 2017, Patz et al. 2014, Haines et al. 2006, McMichael, Haines, and Kovats 1996, Kovats et al. 2003, Confalonieri et al. 2007, Kjellstrom et al. 2009, Kovats et al. 2000, Heaviside 2019, Mitchell et al. 2018, Hajat et al. 2014, Nichols, Lake, and Heaviside 2018, Smith et al. 2014, Nilsson et al. 2012, Woodward et al. 2014, McGushin, Tcholakov, and Hajat 2018).
4 According to Margolis (2013), "Heat stroke is typically divided into two types: 'Exertional Heat Stroke' as the name implies involves strenuous physical activity usually under high temperature conditions to which the person was not acclimated and usually affects healthy older teens and young adults, such as athletes, occupational workers, and soldiers. 'Classical heat stroke', by definition, does not involve exertion and usually affects biologically susceptible individuals, such as infants and young children, the elderly, persons with chronic illness and/or taking medications (prescribed or over-the-counter), as well as person specific actions need to be developed/implemented to reduce potential exposures (to heat, chemical and/or infectious agents) experienced by populations individuals at the local scale."
5 Vardoulakis and Heaviside (2012, 21) describe the urban heat island effect as follows: "Due to the properties of urban building materials, heat sources from human activity, reduced sky-view factor and a lack of moisture in urban areas, temperatures are often a few degrees (around 5–10 °C) higher in urban areas than in surrounding countryside, particularly at night."
6 Homeless people are particularly vulnerable to this health risk since they often have no means to protect themselves from it (Leal Filho et al. 2016b, 5).
7 In public health literature, vulnerable populations are frequently summarized as Young, Old, Pregnant, Immunosuppressed (YOPI).
8 Additionally, a considerable amount of health risks associated resulting from those events can be classified as secondary risks as the following section will explain.
9 This argument only holds true for moderate temperatures and not extreme heat, since extreme temperatures often lead to more time spent indoors.
10 Additionally, diarrheal diseases, such as salmonellosis and campylobacteriosis, are generally more common when temperatures are higher, though variances over place and pathogen may occur (Luber et al. 2014, 226). Accordingly, climate change affects waterborne diseases in multiple ways.
11 The effects of climate change on agriculture will vary from region to region (Campbell-Lendrum et al. 2003, 153).
12 According to the WHO, infectious diseases "are caused by pathogenic microorganisms, such as bacteria, viruses, parasites or fungi; the diseases can be spread, directly or indirectly, from one person to another" (WHO 2020). Zoonotic diseases are infectious diseases of animals that can be transmitted to humans.
13 According to Ojeh and Aworinde (2016, 174), "The optimum temperature for the malaria parasite extrinsic incubation period is between 20 and 27°C, while the maximum temperature for both vector and parasite is 40°C."

14 Dengue fever is primarily transmitted by "Aedes aegypti (now named Stegomyeie aegypti) and secondarily by Aedes (Stegomed) albopictus" (Rom and Pinkerton 2013, 12).

15 Pollen does not necessarily have to be allergic, but an increasing number of people react to it. Exposure to pollen causes mast cells to release histamines into the blood stream resulting in symptoms classified as allergies.

References

Akpinar-Elci, Muge, and Hugh Sealy. 2013. "Climate change and public health in small island states and Caribbean countries." In *Global Climate Change and Public Health*, edited by Kent E. Pinkerton and William N. Rom, 279–292. New York: Springer.

Anderson, Brooke, and Michelle Bell. 2009. "Weather-related mortality: how heat, cold, and heat waves affect mortality in the United States." *Epidemiology* 20 (2): 205–213.

Aparicio-Effen, Marilyn, Ivar Arana, James Aparicio, Cinthya Ramallo, Nelson Bernal, Mauricio Ocampo, and G. J. Nagy. 2016. "Climate change and health vulnerability in Bolivian chaco ecosystems." In *Climate Change and Health: Improving Resilience And Reducing Risks*, edited by Walter Leal Filho, Ulisses de Miranda Azeiteiro and Fátima Alves, 231–259. Cham/Heidelberg: Springer.

Asad, Hina, and David Carpenter. 2018. "Effects of climate change on the spread of zika virus: a public health threat." *Rev. Environ. Health*, 33 (1): 31–42.

Balbus, John, Kristie Ebi, Lauren Finzer, Catherine Malina, Amy Chadwick, Dennis McBride, Michelle Chuk, and Ed Maibach. 2008. *Are We Ready? – Preparing for the Public Health Challenges of Climate Change.* New York: Environmental Defense Fund.

Bambrick, Hilary, and Simon Hales. 2014. "Dengue: distribution and transmission dynamics with climate change." In *Climate Change and Global Health*, edited by Colin Butler, 77–84. Canberra: CABI.

Barbara, Joanna Santa. 2013. "The impact of climate change on human health." In *Impact of Climate Change on Water and Health*, edited by Velma I. Grover, 75–105. Boca Raton: CRC Press.

BBC. 2019a. "European heatwave: France hits record temperature of 45.9°C." *BBC Online*, 28 June 2019. https://www.bbc.com/news/world-europe-48795264.

BBC. 2019b. "Polar vortex death toll rises to 21 as US cold snap continues." *BBC Online*, 1 February 2019. https://www.bbc.com/news/world-us-canada-47088684.

BBC. 2020. "Australia fires: a visual guide to the bushfire crisis." *BBC News*, 31 January 2020. Available at: https://www.bbc.co.uk/news/world-australia-50951043.

Beggs, Paul. 2014. "Impacts of climate change on allergens and allergic diseases: knowledge and highlights from two decades of research." In *Climate Change and Global Health*, edited by Colin Butler, 105–113. Canberra: CABI.

Beggs, Paul. 2016. "Introduction." In *Impacts of Climate Change on Allergens and Allergic Diseases*, edited by Paul Beggs, 1–9. Cambridge: Cambridge University Press.

Berry, Helen Louise, Kathryn Bowen, and Tord Kjellstrom. 2008. "Climate change and mental health: a causal pathways framework." *International Journal of Public Health* 55: 123–132.

Bharath, A. K., and R. J. Turner. 2009. "Impact of climate change on skin cancer." *Journal of the Royal Society of Medicine* 102 (6): 215–218. doi:10.1258/jrsm.2009.080261.

Blume, Joshua, Steven D. Douglas, and Dwight L. Evans. 2011. "Immune suppression and immune activation in depression." *Brain, Behavior, and Immunity* 25 (2): 221–229. doi:10.1016/j.bbi.2010.10.008.

Bowles, D. C., and C. D. Butler. 2014. "Socially, politically and economically mediated health effects of climate change: possible consequences for Africa." *South African Medical Journal = Suid-Afrikaanse tydskrif vir geneeskunde* 104 (8): 585. doi:10.7196/samj.8604.

Bowles, Devin, Mark Braidwood, and Colin Butler. 2014. "Unholy trinity: climate change, conflict and ill health." In *Climate Change and Global Health*, edited by Colin Butler, 144–152. Canberra: CABI.

Bowles, Devin, Rafael Reuveny, and Colin D. Butler. 2014. "Moving to a better life? Climate, migration and population health." In *Climate Change and Global Health*, edited by Colin Butler, 135–143. Canberra: CABI.

Butler, Colin. 2014a. *Climate Change and Global Health*. Canberra: CABI.

Butler, Colin. 2014b. "Famine, hunger, society and climate change." In *Climate Change and Global Health*, edited by Colin Butler, 124–134. Canberra: CABI.

Campbell-Lendrum, Diarmid, C. F. Corvalán, and A. Prüss-Ustün. 2003. "How much disease could climate change cause?" In *Climate Change and Human Health – Risks and Responses*, edited by A. J. McMichael, D. H. Campbell-Lendrum, C. F. Corvalán, K. L. Ebi, A. K. Githeko, J. D. Scheraga and A. Woodward, 133–158. Geneva: World Health Organization.

Cann, K. F., D. Rh Thomas, R. L. Salmon, A. P. Wyn-Jones, and D. Kay. 2013. "Extreme water-related weather events and waterborne disease." *Epidemiology and Infection* 141 (4): 671–686. doi:10.1017/S0950268812001653.

Casas, André Luís Foroni, Gabriella Mendes Dias Santos, Natalia Bíscaro Chiocheti, and Mônica de Andrade. 2016. "Effects of temperature variation on the human cardiovascular system: a systematic review." In *Climate Change and Health: Improving Resilience And Reducing Risks*, edited by Walter Leal Filho, Ulisses de Miranda Azeiteiro and Fátima Alves, 73–87. Cham/Heidelberg: Springer.

Chabejong, Nkwetta Elvis. 2016. "A review on the impact of climate change on food security and malnutrition in the Sahel region of Cameroon." In *Climate Change and Health: Improving Resilience and Reducing Risks*, edited by Walter Leal Filho, Ulisses de Miranda Azeiteiro and Fátima Alves, 133–148. Cham/Heidelberg: Springer.

Charron, D., Fleury, M., Lindsay, R., Ogden, N., and Schuster, C. 2008. "The impacts of climate change on water-, food- and rodent-borne diseases." In *Human Health in a Changing Climate: A Canadian Assessment of Vulnerabilities and Adaptive Capacity*, edited by Jacinthe Séguin, 171–210. Ottawa, Canada: Minister of Health.

Confalonieri, Ulisses, Bettina Menne, Rais Akhtar, Kristie Ebi, Maria Hauengue, R. Sari Kovats, Boris Revich, and Alistair Woodward. 2007. "Human health." In *Climate Change 2007: Impacts, Adaptation and Vulnerability. Contribution of Working Group II to the Fourth Assessment Report of the Intergovernmental Panel on Climate Change*, edited by M. L. Parry, O. F. Canziani, J. P. Palutikof, P. J. van der Linden and C. E. Hanson, 391–431. Cambridge: IPCC.

Cox, Lisa, and Jonathan Watts. 2019. "Australia's extreme heat is sign of things to come, scientists warn." *The Guardian*, 1 February 2019. https://www.theguardian.com/australia-news/2019/feb/01/australia-extreme-heat-sign-of-things-to-come-scientists-warn-climate.

Cromar, Lauren, and Kevin Cromar. 2013. "Dengue fever and climate change." In *Global Climate Change and Public Health*, edited by Kent E. Pinkerton and William N. Rom, 167–191. New York: Springer.

Doherty, Thomas J., and Susan Clayton. 2011. "The psychological impacts of global climate change." *American Psychologist* 66 (4): 265–276. doi:10.1037/a0023141.

Fanzo, Jessica, Claire Davis, Rebecca McLaren, and Jowel Choufani. 2018. "The effect of climate change across food systems: implications for nutrition outcomes." *Global Food Security* 18: 12–19. doi:10.1016/j.gfs.2018.06.001.

Fermeer, Catrin, Art Dewolf, and Gerard Breeman. 2013. "Governance of wicked climate adaptation problems." In *Climate Change Governance*, edited by Jörg Knieling and Walter Leal Filho, 27–39. Berlin/Heidelberg: Springer.

Friel, S., K. Bowen, D. Campbell-Lendrum, H. Frumkin, A. J. McMichael, and K. Rasanathan. 2011. "Climate change, noncommunicable diseases, and development: the relationships and common policy opportunities." *Annual Review of Public Health* 32 (1): 133–147. doi:10.1146/annurev-publhealth-071910-140612.

Gasparrini, Antonio, Yuming Guo, Francesco Sera, Ana Maria Vicedo-Cabrera, Veronika Huber, Shilu Tong, Micheline de Sousa Zanotti Stagliorio Coelho, Paulo Hilario Nascimento Saldiva, Eric Lavigne, Patricia Matus Correa, Nicolas Valdes Ortega, Haidong Kan, Samuel Osorio, Jan Kyselý, Aleš Urban, Jouni J. K. Jaakkola, Niilo R. I. Ryti, Mathilde Pascal, Patrick G. Goodman, Ariana Zeka, Paola Michelozzi, Matteo Scortichini, Masahiro Hashizume, Yasushi Honda, Magali Hurtado-Diaz, Julio Cesar Cruz, Xerxes Seposo, Ho Kim, Aurelio Tobias, Carmen Iñiguez, Bertil Forsberg, Daniel Oudin Åström, Martina S. Ragettli, Yue Leon Guo, Chang-fu Wu, Antonella Zanobetti, Joel Schwartz, Michelle L. Bell, Tran Ngoc Dang, Dung Do Van, Clare Heaviside, Sotiris Vardoulakis, Shakoor Hajat, Andy Haines, and Ben Armstrong. 2017. "Projections of temperature-related excess mortality under climate change scenarios." *The Lancet Planetary Health* 1 (9): e360–e367. doi:10.1016/S2542-5196(17)30156-0.

Gillis, Justin. 2016. "In zika epidemic, a warning on climate change." *New York Times*, 20 February 2016. http://www.nytimes.com/2016/02/21/world/americas/in-zika-epidemic-a-warning-on-climate-change.html?_r=0.

Gislason, Maya K. 2015. "Climate change, health and infectious disease." *Virulence* 6 (6): 539–542.

Grover, Velma I. 2013. "Introduction: impact of climate change on water cycle and health." In *Impact of Climate Change on Water and Health*, edited by Velma I. Grover, 3–29. Boca Raton: CRC Press.

Guy, J., Michael Brottrager, Jesus Crespo Cuaresma, and Raya Muttarak. 2019. "Climate, conflict and forced migration." *Global Environmental Change* 54: 239–249. doi:10.1016/j.gloenvcha.2018.12.003.

Haas, Benjamin. 2018. "South Korean heatwave causes record deaths." *The Guardian*, 9 August 2018. https://www.theguardian.com/world/2018/aug/09/south-korea n-heatwave-causes-record-deaths.

Haines, Andy, R. Sari Kovats, Diarmid Campbell-Lendrum, and Carlos Corvalán. 2006. "Climate change and human health: impacts, vulnerability and public health." *Public Health* 120 (7): 585–596.

Hajat, Shakoor, Sotiris Vardoulakis, Clare Heaviside, and Bernd Eggen. 2014. "Climate change effects on human health: projections of temperature-related mortality for the UK during the 2020s, 2050s and 2080s." *Epidemiol Community Health* 68: 641–648.

Hales, S., S. J. Edwards, and R. S. Kovats. 2003. "Impacts on health of climate extremes." In *Climate Change and Human Health – Risks and Responses*, edited by A. J. McMichael, D. H. Campbell-Lendrum, C. F. Corvalán, K. L. Ebi, A. K. Githeko, J. D. Scheraga and A. Woodward, 79–102. Geneva: World Health Organization.

Hales, Simon, Neil De Wet, John Maindonald, and Alistair Woodward. 2002. "Potential effect of population and climate changes on global distribution of dengue fever: an empirical model." *The Lancet* 360 (9336): 830–834.

Hales, Simon, Sari Kovats, Simon Lloyd, and Diarmid Campbell-Lendrum. 2014. *Quantitative Risk Assessment of the Effects of Climate Change on Selected Causes of Death, 2030s and 2050s*. Geneva: World Health Organization.

Heaviside, Clare. 2019. *Understanding the Impacts of Climate Change on Health to Better Manage Adaptation Action*. Multidisciplinary Digital Publishing Institute.

Heaviside, Clare, Helen Macintyre, and Sotiris Vardoulakis. 2017. "The urban heat island: implications for health in a changing environment." *Current Environmental Health Reports* 4 (3): 296–305. doi:10.1007/s40572-017-0150-3.

Hess, Jeremy J., Josephine N. Malilay, and Alan J. Parkinson. 2008. "Climate change: the importance of place." *American Journal of Preventive Medicine* 35 (5): 468–478.

Hess, Jeremy J., Paul J. Schramm, and George Luber. 2014. "Public health and climate change adaptation at the federal level: one agency's response to Executive Order 13514." *American Journal of Public Health* 104 (3): e22–e30. doi:10.2105/AJPH.2013.301796.

Hess, Jermey J., Gino Marinucci, Paul J. Schramm, Arie Manangan, and George Luber. 2013. "Management of climate change adaptation at the United States centers of disease control and prevention." In *Global Climate Change and Public Health*, edited by Kent E. Pinkerton and William N. Rom, 341–360. New York: Springer.

Hutton, Guy. 2011. "The economics of health and climate change: key evidence for decision making." *Globalization and Health* 7 (18): 1–7.

Kjellstrom, Tord, Sari Kovats, Simon Lloyd, Tom Holt, and Richard Tol. 2009. "The direct impact of climate change on regional labor productivity." *Archives of Environmental & Occupational Health* 64 (4): 217–227.

Kovats, R. Sari, Bettina Menne, Anthony J. McMichael, Carlos Corvalan, and Roberto Bertollini. 2000. *Climate Change and Human Health – Impact and Adaptation*. Geneva/Rome: World Health Organization United Nations.

Kovats, R. S., B. Menne, M. J. Ahern, and J. A. Patz. 2003. "National assessments of health impacts of climate change: a review." In *Climate Change and Human Health – Risks and Responses*, edited by A. J. McMichael, D. H. Campbell-Lendrum, C. F. Corvalán, K. L. Ebi, A. K. Githeko, J. D. Scheraga and A. Woodward, 181–203. Geneva: World Health Organization.

Lake, Iain R., Natalia R. Jones, Maureen Agnew, Clare M. Goodess, Filippo Giorgi, Fabien Solomon, Lynda Hamaoui-Laguel, Robert Vautard, Mikhail A. Semenov, Jonathan Storkev, and Michelle M. Epstein. 2017. "Climate change and future pollen allergy in Europe." *Environmental Health Perspectives* 125 (3): 385–391. doi:10.1289/EHP173.

Lamothe, Dan. 2018. "The new Arctic frontier." 21 November 2018. *The Washington Post*.

Leal Filho, Walter, Ulisses de Miranda Azeiteiro, and Fátima Alves, eds. 2016a. *Climate Change and Health: Improving Resilience and Reducing Risks, Climate Change Management*. Cham/Heidelberg: Springer.

Leal Filho, Walter, Ulisses de Miranda Azeiteiro, and Fátima Alves. 2016b. "Climate change and health: an overview of the issues and needs." In *Climate Change and Health: Improving Resilience and Reducing Risks*, edited by Walter Leal Filho, Ulisses de Miranda Azeiteiro and Fátima Alves, 1–11. Cham/Heidelberg: Springer.

Lloyd, Simon J., R. Sari Kovats, Zaid Chalabi, Sally Brown, and Robert J. Nicholls. 2016. "Modelling the influences of climate change-associated sea-level rise and socioeconomic development on future storm surge mortality." *Climatic Change* 134 (3): 441–455. doi:10.1007/s10584-015-1376-4.

Louis, Michael St., and Jeremy Hess. 2008. "Climate change – impacts on and implications for global health." *American Journal of Preventive Medicine* 35 (5): 527–538.

Luber, George, Kim Knowlton, John Balbus, Howard Frumkin, Mary Hayden, Jeremy Hess, Michael McGeehin, Nicky Sheats, Lorraine Backer, C. Ben Beard, Kristie L. Ebi, Edward Maibach, Richard S. Ostfeld, Christine Wiedinmyer, Emily Zielinski-Gutiérrez, and Lewis Ziska. 2014. "Human health – climate change impacts in the United States: the third national climate assessment." In *Climate Change Impacts in the United States: The Third National Climate Assessment: U.S. Global Change Research Program*, edited by J. M. Melillo, Terese Richmond and G. W. Yohe. Washington DC: US Government Printing Office, 220–256.

Mach, Katharine J., Caroline M. Kraan, W. Neil Adger, Halvard Buhaug, Marshall Burke, James D. Fearon, Christopher B. Field, Cullen S. Hendrix, Jean-Francois Maystadt, John O'Loughlin, Philip Roessler, Jürgen Scheffran, Kenneth A. Schultz, and Nina von Uexkull. 2019. "Climate as a risk factor for armed conflict." *Nature* 571 (7764): 193–197. doi:10.1038/s41586-019-1300-6.

Machalaba, Catherine. 2015. "Climate change and health: transcending silos to find solutions." *Annals of Global Health* 81 (3): 445–458.

Margolis, Helene G. 2013. "Heat waves and rising temperatures: human health impacts and the determinants of vulnerability." In *Global Climate Change and Public Health*, edited by Kent E. Pinkerton and William N. Rom, 85–120. New York: Springer.

McCreesh, Nicky, Grigory Nikulin, and Mark Booth. 2015. "Predicting the effects of climate change on *Schistosoma mansoni* transmission in eastern Africa." *Parasites & Vectors* 8, 4 (2015). doi:10.1186/s13071-014-0617-0.

McGushin, Alice, Yassen Tcholakov, and Shakoor Hajat. 2018. *Climate change and human health: health impacts of warming of 1.5 C and 2 C*. Multidisciplinary Digital Publishing Institute.

McIver, Lachlan, Rokho Kim, Alistair Woodward, Simon Hales, Jeffery Spickett, Dianne Katscherian, Masahiro Hashizume, Yasushi Honda, Ho Kim, Steven Iddings, Jyotishma Naicker, Hilary Bambrick, J. McMichael Anthony, and L. Ebi Kristie. 2016. "Health impacts of climate change in pacific island countries: a regional assessment of vulnerabilities and adaptation priorities." *Environmental Health Perspectives* 124 (11): 1707–1714. doi:10.1289/ehp.1509756.

McMichael, Anthony J. 2013. "Globalization, climate change, and human health." *The New England Journal of Medicine* 368 (14): 1335–1343.

McMichael, Anthony. 2014. "Climate change and global health." In *Climate Change and Global Health*, edited by Colin Butler, 11–20. Canberra: CABI.

McMichael, A. J., A. Haines, and S. Kovats. 1996. *Climate Change and Human Health*. Geneva: World Health Organization United Nations.

McMichael, Anthony, Rosalie Woodruff, and Simon Hales. 2006. "Climate change and human health: present and future risks." *The Lancet* 367: 859–869.

McMichael, Celia, Jon Barnett, and Anthony J. McMichael. 2012. "An ill wind? Climate change, migration, and health." *Environmental Health Perspectives* 120 (5): 646–654. doi:10.1289/ehp.1104375.

Milman, Oliver. 2019. "What is the polar vortex – and how is it linked to climate change?" *The Guardian Online*, 31 January 2019. https://www.theguardian.com/us-news/2019/jan/30/polar-vortex-2019-usa-what-is-it-temperatures-cold-weather-climate-change-explained.

Mitchell, Daniel, Clare Heaviside, Nathalie Schaller, Myles Allen, Kristie L. Ebi, Erich M. Fischer, Antonio Gasparrini, Luke Harrington, Viatcheslav Kharin, Hideo Shiogama, Jana Sillmann, Sebastian Sippel, and Sotiris Vardoulakis. 2018. "Extreme heat-related mortality avoided under Paris Agreement goals." *Nature Climate Change* 8 (7): 551–553. doi:10.1038/s41558-018-0210-1.

Mora, Camilo, Bénédicte Dousset, Iain R. Caldwell, Farrah E. Powell, Rollan C. Geronimo, Coral R. Bielecki, Chelsie W. W. Counsell, Bonnie S. Dietrich, Emily T. Johnston, Leo V. Louis, Matthew P. Lucas, Marie M. McKenzie, Alessandra G. Shea, Han Tseng, Thomas W. Giambelluca, Lisa R. Leon, Ed Hawkins, and Clay Trauernicht. 2017. "Global risk of deadly heat." *Nature Climate Change* 7: 501. doi:10.1038/nclimate3322https://www.nature.com/articles/nclimate3322#supplementary-information.

Morita, Haruka, and Patrick Kinney. 2014. "Wildfires, air pollution, climate change and health." In *Climate Change and Global Health*, edited by Colin Butler, 114–123. Canberra: CABI.

Morton, Adam. 2020. "More than 100 threatened species hit hard by Australian bushfires, pushing many towards extinction." *The Guardian Online*, 20 January 2020. Accessed 24 March 2020. https://www.theguardian.com/environment/2020/jan/20/more-than-100-threatened-species-australian-bushfires-towards-extinction.

Neira, Maria, Roberto Bertollini, Diarmid Campbell-Lendrum, and David L. Heymann. 2008. "The year 2008 – A breakthrough year for health protection from climate change?" *American Journal of Preventive Medicine* 35 (5): 424–425.

Nichols, Gordon, Iain Lake, and Clare Heaviside. 2018. "Climate change and water-related infectious diseases." *Atmosphere* 9 (10). doi:10.3390/atmos9100385.

Nilsson, Maria, Birgitta Evengard, Rainer Sauerborn, and Peter Byass. 2012. "Connecting the global climate change and public health agendas." *PLoS Med* 9 (6): 1–3.

Ogden, Nick. 2014. "Lyme disease and climate change." In *Climate Change and Global Health*, edited by Colin Butler, 85–94. Canberra: CABI.

Ojeh, Vincent Nduka, and Sheyi A. Aworinde. 2016. "Climate variation and challenges of human health in Nigeria: Malaria in perspective." In *Climate Change and Health: Improving Resilience and Reducing Risks*, edited by Walter Leal Filho, Ulisses de Miranda Azeiteiro and Fátima Alves, 171–185. Cham/Heidelberg: Springer.

Okada, H., C. Kuhn, H. Feillet, and J. F. Bach. 2010. "The 'hygiene hypothesis' for autoimmune and allergic diseases: an update." *Clinical & Experimental Immunology* 160 (1): 1–9. doi:10.1111/j.1365-2249.2010.04139.x.

Patz, J. A., A. K. Githeko, J. P. McCarty, S. Hussein, U. Confalonieri, and N. de Wet. 2003. "Climate change and infectious diseases." In *Climate Change and Human Health – Risks and Responses*, edited by A. J. McMichael, D. H. Campbell-Lendrum, C. F. Corvalán, K. L. Ebi, A. K. Githeko, J. D. Scheraga, and A. Woodward, 103–132. Geneva: World Health Organization.

Patz, Jonathan. 2013. "Climate variability and change: food, water, and societal impacts." In *Global Climate Change and Public Health*, edited by Kent E. Pinkerton and William N. Rom, 211–235. New York: Springer.

Patz, Jonathan A. 2018. *Altered Disease Risk From Climate Change*. New York: Springer.

Patz, Jonathan A., and Madeleine C. Thomson. 2018. "Climate change and health: moving from theory to practice." *PLoS Med*, 15 (7), doi:10.1371%2Fjournal. pmed.1002628.

Patz, Jonathan, Carlos Corvalan, Pierre Horwitz, Diarmid Campbell-Lendrum, Nick Watts, MarinaMaiero, SarahOlson, JenniferHales, ClarkMiller, KathrynCampbell, CristinaRomanelli, and David Cooper. 2012. *Our Planet, Our Health, Our Future – Human Health and the Rio Conventions: Biological Diversity, Climate Change and Desertication*. Geneva: World Health Organization.

Patz, Jonathan, Howard Frumkin, Tracey Holloway, Daniel Vimont, and Andrew Haines. 2014. "Climate change challenges and opportunities for global health." *JAMA* 312 (15): 1565–1580.

Pereda, Paula Carvalho, and Denisard Cneio de Oliveira Alves. 2016. "Climate impacts on dengue risk in Brazil: current and future risks." In *Climate Change and Health: Improving Resilience and Reducing Risks*, edited by Walter Leal Filho, Ulisses de Miranda Azeiteiro and Fátima Alves, 201–230. Cham/Heidelberg: Springer.

Rom, William N., and Kent E. Pinkerton. 2013. "Introduction: consequences of global warming to the public's health." In *Global Climate Change and Public Health*, edited by Kent E. Pinkerton and William N. Rom, 1–20. New York: Springer.

Rowlatt, Justin. 2019. "Polar vortex: what role does climate change play?" *BBC Online*, 31 January 2019. https://www.bbc.com/news/world-us-canada-47078054.

Schleussner, C. F., T. K. Lissner, E. M. Fischer, J. Wohland, M. Perrette, A. Golly, J. Rogelj, K. Childers, J. Schewe, K. Frieler, M. Mengel, W. Hare, and M. Schaeffer. 2016. "Differential climate impacts for policy-relevant limits to global warming: the case of 1.5 °C and 2 °C." *Earth Syst. Dynam.* 7 (2): 327–351. doi:10.5194/esd-7-327-2016.

Schnitter, Rebekka, Marielle Verret, Peter Berry, Tanya Chung, Tiam Fook, Simon Hales, Aparna Lal, and Sally Edwards. 2018. "An assessment of climate change and health vulnerability and adaptation in Dominica." *International Journal of Environmental Research and Public Health* 16 (1). doi:10.3390/ijerph16010070.

Semenza, Jan. 2014. "Climate change adaptation to infectious diseases in Europe." In *Climate Change and Global Health*, edited by Colin Butler, 193–205. Canberra: CABI.

Setti, Andréia, Faraoni Freitas, Helena Ribeiro, Edmund Gallo, Fátima Alves, and Ulisses Miranda Azeiteiro. 2016. "Climate change and health: governance mechanisms in traditional communities of Mosaico Bocaina/Brazil." In *Climate Change and Health: Improving Resilience And Reducing Risks*, edited by Walter Leal Filho, Ulisses de Miranda Azeiteiro and Fátima Alves, 329–351. Cham/Heidelberg: Springer.

Shukla, Jyotsana. 2016. "Extreme weather events: addressing the mental health challenges." In *Climate Change and Health: Improving Resilience and Reducing Risks*, edited by Walter Leal Filho, Ulisses de Miranda Azeiteiro and Fátima Alves, 15–27. Cham/Heidelberg: Springer.

Shumake-Guillemot, Joy, Elena Villalobos-Prats, and Diarmid Campbell-Lendrum. 2015. *Operational Framework for Building Climate Resilient Health Systems*. Geneva: United Nations, World Health Organization.

Sim, Walter. 2018. "At least 77 dead in Japan as heatwave pushes temperature to record 41.1 deg C." *The Straits Times*, 23 July 2018. https://www.straitstimes.com/asia/east-asia/japanese-city-sizzles-at-highest-recorded-temperature-of-411-degree-celsius.

Singh, Manpreet, and Mala Rao. 2014. "Climate change and health in South Asian countries." In *Climate Change and Global Health*, edited by Colin Butler, 162–171. Canberra: CABI.

Singh, Reena B., Simon Hales, Neil De Wet, Rishi Raj, Mark Hearnden, and Phil Weinstein. 2001. "The influence of climate variation and change on diarrheal disease in the Pacific Islands." *Environmental Health Perspectives* 109 (2): 155–159.

Smith, Kirk R., Alistair Woodward, Diarmid Campbell-Lendrum, Dave D. Chadee, Yasushi Honda, Qiyong Liu, Jane M. Olwoch, Boris Revich, and Rainer Sauerborn. 2014. "Human health: impacts, adaptation, and co-benefits." In *Climate Change 2014: Impacts, Adaptation, and Vulnerability*, edited by C. B. Field, V. R. Barros, D.J. Dokken, K. J. Mach, M. D. Mastrandrea, T. E. Bilir, M. Chatterjee, K. L. Ebi, Y. O. Estrada, R. C. Genova, B. Girma, E. S. Kissel, A. N. Levy, S. MacCracken, P. R. Mastrandrea and L. L. White, 709–754. Cambridge/New York: Cambridge University Press.

Stott, Peter. 2016. "How climate change affects extreme weather events." *Science* 352 (6293): 1517. doi:10.1126/science.aaf7271.

Taube, Friedel. 2019. "Mozambique after Cyclone Idai: 'some people have not eaten in weeks'." *Deutsche Welle*, 21 April 2019. https://www.dw.com/en/mozambique-after-cyclone-idai-some-people-have-not-eaten-in-weeks/a-48425783.

Vardoulakis, Sotiris, and Clare Heaviside. 2012. *Health Effects of Climate Change in the UK 2012*. London: Health Protection Agency.

Veenema, Tener Goodwin, Clifton Thornton, Roberta Proffitt Lavin, Annah K. Bender, Stella Seal, and Andrew Corley. 2017. "Climate change-related water disasters' impact on population health." *Journal of Nursing Scholarship* 49 (6): 625–634. doi:10.1111/jnu.12328

Vezzulli, Luigi, Rita Colwell, and Carla Pruzzo. 2013. "Ocean warming and spread of pathogenic vibrios in the aquatic environment." *Microbial Ecology* 65: 817–825.

Vicedo-Cabrera, Ana Maria, Yuming Guo, Francesco Sera, Veronika Huber, Carl-Friedrich Schleussner, Dann Mitchell, Shilu Tong, Micheline de Sousa, Zanotti Stagliorio Coelho, Paulo Hilario, Nascimento Saldiva, Eric Lavigne, Patricia Matus, Correa Nicolas, Valdes Ortega, Haidong Kan, Samuel Osorio, Jan Kyselý, Aleš Urban, Jouni J. K. Jaakkola, Niilo R. I. Ryti, Mathilde Pascal, Patrick G. Goodman, Ariana Zeka, Paola Michelozzi, Matteo Scortichini, Masahiro Hashizume, Yasushi Honda, Magali Hurtado-Diaz, Julio Cruz, Xerxes Seposo, Ho Kim, Aurelio Tobias, Carmen Íñiguez, Bertil Forsberg, Daniel Oudin Åström, Martina S. Ragettli, Martin Röösli, Yue Leon Guo, Chang-fu Wu, Antonella Zanobetti, Joel Schwartz, Michelle L. Bell, Tran Ngoc Dang, Dung Do Van, Clare Heaviside, Sotiris Vardoulakis, Shakoor Hajat, Andy Haines, Ben Armstrong, Kristie L. Ebi, and Antonio Gasparrini. 2018. "Temperature-related mortality impacts under and beyond Paris Agreement climate change scenarios." *Climatic Change* 150 (3): 391–402. doi:10.1007/s10584-018-2274-3.

Watts, Nick, Diarmid Campbell-Lendrum, Marina Maiero, Lucia Fernandez Montoya, and Kelly Lao. 2015. *Strengthening Health Resilience to Climate Change –*

Technical Briefing for the World Health Organization Conference on Health and Climate. Geneva: United Nations, World Health Organization.

Woodward, Alistair, Kirk R. Smith, Diarmid Campbell-Lendrum, Dave D. Chadee, Yasushi Honda, Qiyong Liu, Jane Olwoch, Boris Revich, Rainer Sauerborn, and Zoë Chafe. 2014. "Climate change and health: on the latest IPCC report." *The Lancet* 383 (9924): 1185–1189.

World Health Organization (WHO). 2017. *Dengue Fever – Sri Lanka.* Geneva: United Nations World Health Organization.

World Health Organization, United Nations. (WHO). 2003. *Climate Change and Human Health – Risks and Responses – Summary.* Geneva: WHO.

World Health Organization, United Nations (WHO/UN). 2009. *Climate Change Is Affecting Our Health – Something Should Be Done Now.* Geneva: WHO.

World Health Organization, United Nations (WHO/UN). 2020. "Infectious diseases." Accessed 30 December 2020. http://www.emro.who.int/health-topics/infectious-diseases/index.html.

World Health Organization, United Nations. (WHO). n.d.-a. *Gender, Climate Change and Health.* Geneva: WHO.

World Health Organization, United Nations. (WHO). n.d.-b. "Water, health and ecosystems." Accessed 20 November 2016. http://www.who.int/heli/risks/water/water/en/.

World Health Organization, United Nations, and United Nations World Meteorological Organization. (WHO/WMO). 2012. *Atlas of Health and Climate.* Geneva: WHO.

Ziska, Lewis, and Paul Beggs. 2012. "Anthropogenic climate change and allergen exposure: the role of plant biology." *Journal of Allergy and Clinical Immunology* 129: 27–32.

Ziska, Lewis H. 2016. "Impacts of climate change on allergen seasonality." In *Impacts of Climate Change on Allergens and Allergic Diseases*, edited by Paul Beggs, 92–112. Cambridge: Cambridge University Press.

2 Politics and the nexus between climate change and health

Politics plays a crucial role with regards to the climate change and health nexus since governments have significant leverage in terms of developing, facilitating, and implementing measures to reduce climate change related health risks. The most effective form of reducing climate change related health risks is climate change mitigation. Nothing helps more to reduce the consequences of climate change than mitigating climate change itself. At the same time, many mitigation actions include a high number of health co-benefits (Smith et al. 2014, 714). However, climate change is no longer simply a challenge for the future, but it has already started, and the effects can be felt across the planet. Even in the most optimistic IPCC climate change scenario, the earth's temperature will be warming by at least 1.5 to 2 °C, which will already result in numerous additional health risks (IPCC 2014, 59). Therefore, it is clear that, on the one hand, mitigation efforts need to be strengthened and, on the other hand, additional measures need to be undertaken to reduce climate change related health risks.

Accordingly, adaptation to climate change needs to play a pivotal role when it comes to preparing for the future. However, different actors have different interpretations of adaptation. Therefore, this chapter aims to provide a clear definition of adaptation and adaptation to climate change related health risks and discuss why health adaptation constitutes a complex challenge with high levels of uncertainty.

What is adaptation?

According to the United Nations Framework Convention on Climate Change (UNFCCC), adaptation

> refers to adjustments in ecological, social, or economic systems in response to actual or expected climatic stimuli and their effects or impacts. It refers to changes in processes, practices, and structures to moderate potential damages or to benefit from opportunities associated with climate change.

> (UNFCCC 2014)

These changes include: "observation; assessment of climate impacts and vulnerability; planning; implementation; monitoring and evaluation of adaptation actions" (UNFCCC 2014). Although the UNFCCC's definition of climate change adaptation is internationally recognized, it contains a number of pitfalls which need to be clarified before delving into research on this subject.

Most importantly, and frequently also discussed under the term "dependent variable problem," it is not always clear whether adaptation measures are actually intended by the implementing actors (Dupuis and Biesbroek 2013). For example, if someone buys an air-conditioning unit because it is warm outside, does that mean that he or she is adapting to climate change? On the state level, similar issues may arise. If countries have been prone to flood risks for centuries and they are now investing more into flood protection, while at the same time climate change exacerbates the existing risks, does that mean that the states are adapting to climate change? To better compare the measures and identify what and how much states do to respond to climate change related risks, we consequently require a refined definition. Accordingly, the UNFCCC's definition will be modified to the following definition of adaptation to climate change related health risks.

Adaptation to climate change related health risks refers to intended structural and procedural adjustments in ecological, social, political or economic systems to moderate or prevent actual or expected negative effects of climate change on health. The actions need to be explicitly communicated as a response to climate change and may address primary, secondary, or tertiary climate change related health risks.

Moreover, adaptation and adaptive capacity are frequently mixed and used synonymously, although they represent two separate categories. Whilst adaptation describes the actions states take to respond to climate change and prevent potential damage, adaptive capacity describes the essential determinants of states' ability to adapt to climate change, such as their existing health systems or overall economic performance (Kovats, Ebi, and Menne 2003, 196). Consequently, the two categories differ with regard to their functional dimension: adaptation refers to the action states take, and adaptive capacity describes their ability to act.

A special challenge

Adaptation to climate change is a very difficult task because of at least two reasons: (a) it is directed towards the future and the future is by definition uncertain, which poses great limits on preparing and planning for it, and (b) it is very complex, highly resistant to solutions, interconnected with other problems, and therefore often described as a "wicked problem" (Fermeer, Dewolf, and Breeman 2013, 27). The adjective wicked is used to distinguish such a problem from other, tameable problems and has nothing to do with witches or wizards (Fermeer et al. 2013, 28–29, Rittel and Webber 1973, 160). According to Rittel and Webber (1973, 160), who introduced the concept of wicked problems, these problems are very hard to identify, can be the symptom of another problem, and can be very frustrating for a great number of different actors, because they never know for sure whether they are actually solving the issue.

Climate change adaptation has been called a "wicked problem par excellence," because it is very fragmented, there is a lack of structure, and the topic is very knowledge-intensive, while a certain degree of uncertainty will always persist since the problem is directed towards the future (Fermeer et al. 2013, 27). It is very difficult to find optimal solutions for such problems since it is very hard to define when adaptation actually succeeds, who benefits from the adaptation measures, and who might actually experience negative consequences. Comparatively simple solutions, such as spraying pesticides to reduce the prevalence of vector borne diseases, often are unsustainable and exacerbate other problems. Rittel and Webber (1973, 162) explain: "In solving a chess problem or a mathematical equation, the problem-solver knows when he has done his job. There are criteria that tell when *the* or *a* solution has been found." When it comes to adaptation however, they argue, a planning problem exists since new challenges can always arise and people can never be sure whether the strategies and actions they implement are sufficient. Due to the complexity and high level of uncertainty, solutions can never be ultimate nor true or false (Rittel and Webber 1973, 162).

For many states, adaptation to climate change related health risks constitutes an even greater and more complex challenge since the levels of uncertainty are higher than for other adaptation challenges, and as the trichotomy of primary, secondary, and tertiary risks shows, health is highly interconnected with other risks (McMichael 2013). A number of researchers, national and regional governments, international organizations, and civil society representatives have come up with concrete proposals on what adaptation to climate change related health risks can look like (Watts et al. 2018).[1] Whilst some countries, such as Germany, were rather early with developing their national and regional adaptation strategies, other countries are still catching up (BMUB 2008, Biesbroek et al. 2010). Some countries have established specific institutions within ministries or governmental agencies to deal with climate change and health, whereas others are chronically understaffed and have not managed to conduct a proper vulnerability assessment yet.[2]

The complexity of health adaptation to climate change is powerfully illustrated by the various sustainability challenges that potential solutions imply. Whilst some adaptation initiatives may seem straightforward and clear, they often come with a number of unintended side effects. Adaptation actions aimed at responding to heatwaves include, among other things, installing air-conditioners and fans, opening public cooling centers, moving to safer or cooler areas during extreme heat or cold, re-scheduling work patterns, and changing clothing habits (Füssel, Klein, and Ebi 2006, 43, Klein Rosenthal and Brechwalt 2013, 210). Although air-conditioning definitely helps to protect people against extreme heat, it comes with very high energy consumption. More air-conditioning thus usually leads to higher energy demand, which is met through power plants that run on fossil fuels and thus exacerbate the climate change problem (Pierre-Louis 2018). Sustainable solutions are usually more complex (e.g. better insulation and infrastructure, heat warning systems, etc.) and therefore often take more time and political will to be implemented than simple solutions (Füssel et al. 2006, 43, NDRC 2017).

At the same time, if planned and implemented properly, adaptation can be sustainable and help to meet mitigation goals. As a measure to reduce health effects of air pollution, which are exacerbated by climate change, some countries, regions, and cities have started to develop more sustainable forms of transportation and energy production, improving access to green spaces and limiting traffic in cities (Klausbruckner et al. 2016, Watts et al. 2015, 1862). Similarly, more sophisticated agricultural and water production technology can not only help to reduce the effects of climate change on health risks associated with food and water, but also to transition to a more sustainable use of natural resources (Smith et al. 2014, 73). Moreover, recycling wastewater and improving technology to efficiently and effectively use alternative sources of water can be useful to improve resilience to droughts and desertification and at the same time save valuable resources for future generations (Campbell-Lendrum et al. 2009, 1664).

As the various challenges and the complex nature of health adaptation to climate change demonstrate, effective adaptation requires a thorough understanding of the specific risks. Since a great variety of factors needs to be considered, inter-, cross- and trans-disciplinary research on climate change and health is essential (Heaviside 2019). Similarly, an increasing number of research projects in the field of climate change can no longer rely on singular research methods but requires the combination of quantitative and qualitative traditions in cohesive mixed methods design.

Notes

1 To better compare the health adaptation initiatives of different countries, Lesnikowski et al. (2013, 1155) distinguish into different levels of adaptation (recognition, groundwork, and adaptation) and different types of adaptation, recognition-level, groundwork-level, and adaptation-level actions.
2 Detailed information on successful examples and countries that experience major challenges in terms of health adaptation to climate change will follow in Parts II and III.

References

Biesbroek, G. Robbert, Rob J. Swart, Timothy R. Carter, Caroline Cowan, Thomas Henrichs, Hanna Mela, Michael D. Morecroft, and Daniela Rey. 2010. "Europe adapts to climate change: comparing National Adaptation Strategies." *Global Environmental Change* 20 (3): 440–450. doi:10.1016/j.gloenvcha.2010.03.005.

BMUB. 2008. *Deutsche Anpassungsstrategie an den Klimawandel*, edited by Naturschutz Bundesministerium für Umwelt, Bau und Reaktorsicherheit.

Campbell-Lendrum, Diarmid, Roberto Bertollini, MariaNeira, KristieEbi, and Anthony McMichael. 2009. "Health and climate change: a roadmap for applied research." *The Lancet* 373: 1663–1665.

Dupuis, J., and G. R. Biesbroek. 2013. "Comparing apples and oranges: the dependent variable problem in comparing and evaluating climate change adaptation policies." *Global Environmental Change: Human and Policy Dimensions* 23 (6): 1476–1487. doi:10.1016/j.gloenvcha.2013.07.022.

Fermeer, Catrin, Art Dewolf, and Gerard Breeman. 2013. "Governance of wicked climate adaptation problems." In *Climate Change Governance*, edited by Jörg Knieling and Walter Leal Filho, 27–39. Berlin/Heidelberg.

Füssel, Hans-Martin, Richard J. T. Klein, and Kristie L. Ebi. 2006. "Adaptation assessment for public health." In *Climate Change and Adaptation Strategies for Human Health*, edited by Bettina Menne and Kristie L. Ebi, 41–62. Darmstadt.

Heaviside, Clare. 2019. "Understanding the impacts of climate change on health to better manage adaptation action." *Atmosphere* 10 (3). doi:10.3390/atmos10030119.

IPCC. 2014. *Climate Change 2014: Synthesis Report. Contribution of Working Groups I, II and III to the Fifth Assessment Report of the Intergovernmental Panel on Climate Change*. Edited by R. K. Pachauri and L. A. Meyer (eds), Core Writing Team. Geneva: IPCC.

Klausbruckner, Carmen, Harold Annegarn, Lucas Hennemann, and Peter Rafaj. 2016. "A policy review of synergies and trade-offs in South African climate change mitigation and air pollution control strategies." *Environmental Science & Policy* 57: 70–78.

Klein Rosenthal, Joyce, and Dona Brechwalt. 2013. "Climate adaptive planning for preventing heat-related health impacts in New York City." In *Climate Change Governance*, edited by Jörg Knieling and Walter Leal Filho, 205–225. Berlin/Heidelberg.

Kovats, Sari, Kristie L. Ebi, and Bettina Menne, eds. 2003. *Methods of Assessing Human Health Vulnerability and Public Health Adaptation to Climate Change, Health and Global Environment Change*. Copenhagen: World Health Organization, Regional Office for Europe [u.a.].

Lesnikowski, A. C., J. D. Ford, L. Berrang-Ford, M. Barrera, P. Berry, J. Henderson, and S. J. Heymann. 2013. "National-level factors affecting planned, public adaptation to health impacts of climate change." *Global Environmental Change* 23 (5): 1153–1163. doi:10.1016/j.gloenvcha.2013.04.008.

McMichael, Anthony J. 2013. "Globalization, climate change, and human health." *The New England Journal of Medicine* 368 (14): 1335.

NDRC. 2017. *China's Policies and Actions for Addressing Climate Change*. Beijing: National Development and Reform Commission.

Pierre-Louis, Kendra. 2018. "The world wants air-conditioning. That could warm the world." *New York Times Online*, 15 May 2018. https://www.nytimes.com/2018/05/15/climate/air-conditioning.html.

Rittel, Horst W. J., and Melvin M. Webber. 1973. "Dilemmas in a general theory of planning." *Policy Sciences* 4 (2): 155–169. doi:10.1007/BF01405730.

Smith, Kirk R., Alistair Woodward, Diarmid Campbell-Lendrum, Dave D. Chadee, Yasushi Honda, Qiyong Liu, Jane M. Olwoch, Boris Revich, and Rainer Sauerborn. 2014. "Human health: impacts, adaptation, and co-benefits." In *Climate Change 2014: Impacts, Adaptation, and Vulnerability*, edited by C. B. Field, V. R. Barros, D. J. Dokken, K. J. Mach, M. D. Mastrandrea, T. E. Bilir, M. Chatterjee, K. L. Ebi, Y. O. Estrada, R. C. Genova, B. Girma, E. S. Kissel, A. N. Levy, S. MacCracken, P. R. Mastrandrea and L. L. White, 709–754. Cambridge/New York: Cambridge University Press.

UNFCCC. 2014. "FOCUS: adaptation." Accessed 26 August 2016. http://unfccc.int/focus/adaptation/items/6999.php.

Watts, Nick, Markus Amann, Nigel Arnell, Sonja Ayeb-Karlsson, Kristine Belesova, Helen Berry, Timothy Bouley, Maxwell Boykoff, Peter Byass, Wenjia Cai, Diarmid Campbell-Lendrum, Jonathan Chambers, Meaghan Daly, Niheer Dasandi, Michael Davies, Anneliese Depoux, Paula Dominguez-Salas, Paul Drummond, Kristie L. Ebi, Paul Ekins, Lucia Fernandez Montoya, Helen Fischer, Lucien Georgeson, Delia Grace, Hilary Graham, Ian Slava Hamilton, Stella Hartinger, Jeremy Hess, Ilan Kelman, Gregor Kiesewetter, Tord Kjellstrom, Dominic Kniveton, Bruno Lemke, Lu Liang, Melissa Lott, Rachel Lowe, Maquins Odhiambo Sewe, Jaime Martinez-Urtaza, Mark Maslin, Lucy McAllister, Jankin Mikhaylov, James Milner, Maziar Moradi-Lakeh, Karyn Morrissey, Kris Murray, Maria Nilsson, Tara Neville, Tadj Oreszczyn, Fereidoon Owfi, Olivia Pearman, David Pencheon, Steve Pye, Mahnaz Rabbaniha, Elizabeth Robinson, Joacim Rocklöv, Olivia Saxer, Stefanie Schütte, Jan C. Semenza, Joy Shumake-Guillemot, Rebecca Steinbach, Meisam Tabatabaei, Julia Tomei, Joaquin Trinanes, Nicola Wheeler, Paul Wilkinson, Peng Gong, Hugh Montgomery, and Anthony Costello. 2018. "The 2018 report of the *Lancet* countdown on health and climate change: shaping the health of nations for centuries to come." *The Lancet* 392 (10163): 2479–2514. doi:10.1016/S0140-6736(18)32594-7.

Watts, Nick, Diarmid Campbell-Lendrum, Marina Maiero, Lucia Fernandez Montoya, and Kelly Lao. 2015. *Strengthening Health Resilience to Climate Change – Technical Briefing for the World Health Organization Conference on Health and Climate*. Geneva: United Nations, World Health Organization.

3   The explanatory model

Despite the growing number of research projects on the nexus between climate change and health, some significant research gaps exist that this study seeks to fill. To date no systematic comparison of the adaptation measures to climate change related health risks of a high number of states exists. Although some comparative studies on health adaptation to climate change have recently emerged, none of them provides a global overview (Paterson et al. 2012, Lesnikowski, Ford, Berrang-Ford, Barrera, Berry, et al. 2013, Austin et al. 2016). In a similar vein, no global index exists that would show how all countries in the world adapt to climate change related health risks. Moreover, none of the existing studies systematically analyzes the causal relations behind states' decisions regarding health adaptation to climate change from a political science perspective. Although some studies on the drivers of and barriers to climate change adaptation have been published in recent years, none of them has a strong political science background and not one analyzes the risk perceptions and causal frameworks behind governmental decision-making in this field through a combination of quantitative and qualitative methods (Berrang-Ford et al. 2014, Biesbroek et al. 2010, Lesnikowski et al. 2011, Lesnikowski, Ford, Berrang-Ford, Barrera, Berry, et al. 2013).

To develop a better understanding of health adaptation to climate change, this study seeks to answer the following questions:

a   How do states perceive and adapt to climate change related health risks?
b   Which factors influence states' perception of and measures to adapt to climate change related health risks?

To answer the research questions and to fill the existing research gap in the field of climate change and health, this thesis seeks to achieve the following:

a   Conduct interdisciplinary research on climate change and health that incorporates perspectives from political science and public health literature

b   Establish a global index that compares the health adaptation initiatives of a high number of states
c   Develop an innovative mixed methods design that connects quantitative and qualitative research traditions to contribute to a better understanding of the drivers of and barriers to health adaptation to climate change from a political science perspective

What we know

Despite the rapidly growing number of publications on different health effects of climate change, not many studies exist that focus on tracking and analyzing health adaptation to climate change. Whilst first studies on smaller groups of states and single case studies have emerged, not much is known about how a large number of states perceives and adapts to such risks and why states differ in their responses to this long-term policy problem (Austin et al. 2016, Biesbroek et al. 2010, Lesnikowski, Ford, Berrang-Ford, Barrera, Berry, et al. 2013).

Nonetheless, some significant first studies on the topic have been published, with many of them originating from the TRAC[3] – Tracking Adaptation to Climate Change Consortium.[1] Biesbroek et al. (2010) were among the first authors to compare national adaptation strategies of states to climate change. In their analysis of seven European Union (EU) Member States, they looked for factors that influence and facilitate the development of national adaptation strategies (Biesbroek et al. 2010). Among the key drivers of adaptation, they identified the "scientific and technical support needed for the development and implementation of such a strategy" and "the role of the strategy in information, communication and awareness-raising of the adaptation issue" (Biesbroek et al. 2010, 440). They further wanted to discover forms of multi-level governance which were used to implement the respective adaptation actions, and to find out "how the strategy addresses integration and coordination with other policy domains" (Biesbroek et al. 2010, 440). However, the study focuses more on climate change adaptation in general and does not provide a detailed analysis of health related adaptation measures (Biesbroek et al. 2010, 443). Moreover, it only takes European countries into account and thus does not provide a global overview. Nonetheless, it constitutes an important contribution to research on the topic since it helps to better understand the potential drivers of and barriers to adaptation to climate change, which is essential for all countries across the globe to better prepare for and respond to the effects of climate change.

Juhola and Westerhoff (2011) followed a similar track by identifying challenges to adaptation to climate change. However, they focused on only two European countries (Finland and Italy) and paid particular attention to the governance of adaptation. They came to the conclusion that adaptation governance is mainly taking place at formal institutions as well as networks of different actors on multiple scales and levels (Juhola and Westerhoff 2011, 239).

In contrast to the study of Biesbroek et al. (2010), Juhola and Westerhoff (2011) do not mention adaptation to health risks, yet they shed light on how adaptation is governed and which role informal networks play in this regard.

Huang et al. (2011) followed a different approach in their quest to learn more about the drivers of and barriers to health adaptation. Unlike their predecessors, they did not analyze primary documents, but conducted an extensive literature review and noted that a combination of factors, including uncertainties of future climate change, financial, technological, and institutional barriers, and social and individual factors may constrain health adaptation to climate change (Huang et al. 2011).

Berrang-Ford, Ford, and Paterson (2011), on the other hand, chose the path of comparing national adaptation strategies and plans and released one of the first significant global comparisons of national-level adaptation measures. In their groundbreaking study, they found out that the adaptation profiles of countries differ greatly between high- and low-income countries (Berrang-Ford et al. 2011, 31). Whilst their study does not focus on health adaptation to climate change, health is included in the overall adaptation assessments, and the health community has learned a lot from their publication.

In the same year, Lesnikowski et al. (2011) published one of the most significant studies on adaptation to the health impacts of climate change so far. They systematically compared the Fifth National Communications (NCs) of 38 Annex I parties to the UNFCCC in order to analyze whether and which type of adaptation actions to climate change related health risks were taken by these states (Lesnikowski et al. 2011). In total, they analyzed 1,912 initiatives and found that 80 percent of the identified actions were groundwork-level actions[2] and only 20 percent constituted tangible actions (Lesnikowski et al. 2011). While the study was very important for research on this subject, since it was the first significant project that systematically compared adaptation to climate change related health risks, it still comes with some pitfalls. Furthermore, the study takes only 38 states into account, which limits the potential for drawing general conclusions for the actions of different regime types or discovering other major general trends in the international system. Nonetheless, the study constitutes a valuable contribution to research in this sector – not only because of its empirical results, but also because of the analytical framework it developed for conducting research on this matter (Lesnikowski et al. 2011, 5).

The year 2013 was of particular importance for research on adaptation to climate change since several groundbreaking studies on this subject were published. Building on their previous work, Lesnikowski, Ford, Berrang-Ford, Barrera, and Heymann (2013, 1) presented the results from a substantially more comprehensive study, which encompassed 104 adaptation initiatives. Similar to their first study, the authors analyzed the NCs by the parties to the convention and took data from the UNFCCC website into account which was published until July 2012 (Lesnikowski, Ford,

Berrang-Ford, Barrera, and Heymann 2013, 4). The researchers developed 12 indicators with which they aimed to identify and characterize how countries adapt to climate change (Lesnikowski, Ford, Berrang-Ford, Barrera, and Heymann 2013, 6). Compared with previous studies, the article was of special importance for research on adaptation to climate change since it constitutes the first ever comparison of adaptation actions of 117 states with a sophisticated and comprehensive research approach. Nonetheless, it contains similar pitfalls as previous studies: it focuses on national adaptation actions and does not really take into account what happened on the subnational level (Lesnikowski, Ford, Berrang-Ford, Barrera, and Heymann 2013, 10). Additionally, the study deals with adaptation to climate change in general and thus, despite including health as one of the analyzed sectors, is not capable of providing in-depth and comprehensive information on the levels of health adaptation across all states. Moreover, despite great progress on the number of analyzed states, it does not take all United Nations (UN) Member States into account.

In the same year, Panic and Ford (2013) published a paper in which they reviewed national-level adaptation to infectious diseases in 14 countries of the Organisation for Economic Co-operation and Development (OECD). Contrary to Lesnikowski, Ford, Berrang-Ford, Barrera, Berry, et al. (2013), they did not use NCs as their database, but focused on evidence from peer-reviewed literature (Panic and Ford 2013, 7083). They specifically tried to "identify the types of adaptations proposed in adaptation plans of national agencies and governments and examine whether there are gaps in current public health planning by comparing with recommendations for infectious disease adaptation in the peer reviewed literature" (Panic and Ford 2013, 7083). They developed a useful analytical framework for categorizing different infectious diseases and gained important empirical results (Panic and Ford 2013, 7086–7097). However, similar to previous studies, they only focus on the national level and they specifically put infectious diseases into the center of their research, thus leaving out all other climate change related health risks.

In 2014, Berrang-Ford et al. (2014) tried to answer the question "What drives national adaptation?" through a comparison of national level adaptation measures of 117 states. They identified institutional capacity, and particularly good governance, as the key driving force for successful adaptation (Berrang-Ford et al. 2014, 448). The NCs between 2008 and 2012 (NC5) served as the major database for the study (Berrang-Ford et al. 2014, 442). The article is of special importance for research on adaptation to climate change since it not only describes the performance of states, but also seeks to answer why their performance differs. At the same time, however, it deals with adaptation in general and does not focus on health. Moreover, although with 117 countries the database is significantly larger than other studies on adaptation, it still does not cover all UN Member States.

In the same year, Massey et al. (2014) published an article with a clear focus on adaptation in Europe. They analyzed differences and similarities concerning the adoption and diffusion of adaptation policies across 29 European countries (Massey et al. 2014, 434). Their key interest consisted in discovering the major driving forces of adaptation to climate change (Massey et al. 2014, 434). They came to the conclusion that internal factors, rather than external ones, are the major reasons for states' progress on adaptation (Massey et al. 2014, 434). Similar to previous studies, they do not have a specific focus on health and adhere to a rather small selection of cases (Massey et al. 2014, 434).

Building on their previous work, Lesnikowski et al. (2015) published an article in 2015 on "national-level progress on adaptation" in which they tracked the adaptation efforts of 41 high-income countries between 2010 and 2014. Their major research goal was to investigate whether states had been making progress on adaptation to climate change compared with the year 2010 (Lesnikowski et al. 2015, 261). As in their previous works, their database mostly consisted of the NCs to the UNFCCC (in this case NC6) (Lesnikowski et al. 2015, 261). The study was of great importance for research on the subject, mainly due to its ability to track progress on adaptation as opposed to focusing on the status quo. It did not, however, focus on health and had a rather restricted case selection.

Another important year for research on adaptation to climate change, and especially to climate change related health risks, was 2016. Araos et al. published two articles on adaptation to climate change in large cities (Araos, Austin, et al. 2016, Araos, Berrang-Ford, et al. 2016). In "Public Health Adaptation to Climate Change in Large Cities: A Global Baseline," the former group of authors compared the health adaptation initiatives of 401 urban areas around the world that have more than one million inhabitants (Araos, Austin, et al. 2016, 53). According to their findings, only 10 percent of the analyzed urban areas have any health adaptation actions at all (Araos, Austin, et al. 2016, 53). The research team collected the data through google searches of the respective cities and the city websites (Araos, Austin, et al. 2016, 56). The second study by the research group around Araos, Berrang-Ford, et al. (2016) is called "Climate Change Adaptation Planning in Large Cities: A Systematic Global Assessment." It is similar to the previous one since it also compares the adaptation actions of different cities across the globe. However, in their article from 2016, the focus does not rest on health adaptation but rather adaptation in general (Araos, Berrang-Ford, et al. 2016). It concludes that from 401 global cities with more than one million inhabitants only 61 cities (15 percent) have any adaptation activities whatsoever and 73 cities (18 percent) plan on developing adaptation strategies in the future (Araos, Berrang-Ford, et al. 2016, 1). Most cities with extensive adaptation actions are "large cities located in high-income countries in North America, Europe, and Oceania, and are adapting to a variety of expected impacts" (Araos, Berrang-Ford, et al. 2016, 1). Both studies brought new insights into research on climate change and health and shed light on a thus far less researched aspect of adaptation, namely the actions taking place in large cities. They further enhanced and expanded the available tools for tracking adaptation since they worked on the instruments to classify adaptation into different categories and to collect data from public websites (Araos, Berrang-Ford, et al. 2016, 2). However, the article does not answer key questions concerning health adaptation of a large number of states and does not take the influence of the international level into account.

One of the most relevant articles in 2016 was published by Austin et al. (2016) under the title "Public Health Adaptation to Climate Change in OECD Countries." The research team analyzed the national-level health adaptation to climate change in ten OECD countries "using publicly available information in government documents and websites" (Austin et al. 2016, 5). In contrast to previous studies on health adaptation to climate change, Austin et al. (2016, 6) did not only count the number of adaptation initiatives in the respective country but categorized them more clearly and identified a number of significant aspects in the adaptation initiatives, such as the role of vulnerable groups. Additionally, the authors were able to show that the high levels of adaptive capacity of many high-income countries often do not translate into actual adaptation actions (Austin et al. 2016, 11). The article is a major contribution to research on public health adaptation to climate change since, in addition to the empirical findings, it further helps to develop instruments and categories for health adaptation tracking. At the same time, it does not provide a comparison of a large number of states but focuses only on ten cases. It further lacks the ability to take subnational and international actions into account since it focuses solely on national-level adaptation.

Focusing on low- and middle-income countries, Ebi and Barrio (2017) evaluated multi-national health adaptation programs to identify lessons learned and evaluate the impact of international organizations, such as the WHO or the United Nations Development Programme (UNDP). Whilst their study offers neither a comparison of the current level of health adaptation in different countries across the globe nor an analysis of various factors that may influence health adaptation policies, it provides valuable insights into the specific details of health adaptation projects and the roles of specific stakeholders within countries and their cooperation with international actors (Ebi and Barrio 2017, 2–6).

Berry et al. (2018) complemented the data from the WHO's Climate and Health Country Profile project with additional Internet searches to publish a review on international progress on climate change and health vulnerability and adaptation assessments. They show that the number of vulnerability and adaptation assessments has risen continuously in recent years, but, according to their findings, only 92 countries have so far completed national vulnerability and adaptation assessments (Berry et al. 2018, 7). Although the study of Berry et al. (2018) constitutes an important addition to the literature on adaptation to climate change related health risks, it comes with a number of pitfalls that require further research in the field. First, the database is rather limited since it rests on WHO country reports on climate change and health and, as several country experts have stated in the interviews for this thesis, the WHO surveys and assessments were often not able to provide a complete overview of the health adaptation policies. They focused on a number of sectors that structured the assessments and surveys rather than analyzing all official adaptation documents in detail (Expert Interview, November 2018). As a consequence, although the assessment provides a solid overview, it struggles with explaining the details of national-level health adaptation. Moreover, despite briefly mentioning the influence of international organizations, such as the WHO, on the development of vulnerability and adaptation assessments in developing countries, the article does not really delve into the drivers of and barriers to health adaptation.

In addition to the increasing number of comparative studies on climate change adaptation, literature on specific regions, such as the Pacific Islands (e.g. Kim, Costello, and Campbell-Lendrum 2015, McIver et al. 2016), comparisons of small numbers of states (Austin et al. 2018), single countries (Hess, Schramm, and Luber 2014, Rakotoarison et al. 2018, Schnitter et al. 2018), or the city and local level (Austin et al. 2019, Tompkins et al. 2010, Wise et al. 2014) has started to grow. Moreover, some conceptual articles on climate change adaptation have been released by Dupuis and Biesbroek (2013), Ford et al. (2013), and Ford and Berrang-Ford (2016). Ford et al. (2013) seek to improve instruments for tracking adaptation. Through reviewing and discussing popular measures for tracking adaptation to climate change and adding insights from their own research projects, they present a set of different approaches that can be used for tracking adaptation, thus facilitating the process for future studies on this subject. Dupuis and Biesbroek (2013) discuss the difficulties of comparing the adaptation policies of different countries with one another and provide a number of potential solutions for some of the most common challenges. Ford and Berrang-Ford (2016) discuss quality standards of tracking adaptation to climate change and summarize them in the 4Cs: "consistency, comparability, comprehensiveness, coherency." The authors believe that successful adaptation tracking projects require "a consistent and operational conceptualization of adaptation," need to "focus on comparable units of analysis," shall "use and develop comprehensive datasets on adaptation action," and need to be coherent with their interpretation of "real adaptation" (Ford and Berrang-Ford 2016, 839).

Research gap

Despite remarkable progress in research on tracking, comparing, and understanding health adaptation to climate change, a large research gap exists. To my knowledge, there are no large-scale mixed methods studies on health adaptation so far and the number of political science studies on the subject is very limited. Importantly, none of the existing studies in the field achieves the following:

- a detailed, transparent, structured, and in-depth comparison of the health adaptation measures of all 193 UN Member States that captures the full scale and level of adaptation
- a mixed-methods research design that allows to test existing theoretical assumptions and develop a new thorough understanding of the variance between the health adaptation performances of different countries

Therefore, this study seeks to contribute to filling the research gap with a mixed methods research design that a) consists out of the creation of a new climate change and health adaptation index, a statistical analysis of potential drivers of and barriers to health adaptation to climate change, and several case studies to test the robustness of the index and the findings of the statistical analysis and b) helps to build new theoretical concepts that shall help to understand how and why different states adapt to climate change related health risks, and c) conducts an additional statistical test of these theoretical assumptions.

Why do states adapt differently?

As the Climate Change and Health Adaptation Index (CHAIn) shows, how states respond to climate change related health risks varies to a great extent.[3] Whilst some countries, such as the UK or the Republic of Korea (RoK), recognize a wide range of climate change related health risks and have implemented a variety of health adaptation measures, other states remain almost completely inactive and do not acknowledge the effects of climate change on health in their national documents. The pressing question, therefore, is how such variations can be explained. Following a mixed methods design,[4] the theoretical has originally been divided into three segments: 1. the deduction of several independent variables based on current academic literature on the drivers of and barriers to adaptation to climate change; 2. the inductive (theory building) analysis of potential additional factors that may influence the adaptation initiatives of states in response to climate change related health risks; 3. a concluding theoretical model of the drivers of and barriers to health adaptation to climate change after the factors from the inductive part have been tested quantitatively again.

To improve the readability of the results, the theoretical model has been streamlined and combined into the following sections. It now consists of hypotheses derived from existing theories, as well as findings from the case studies. The original research design, including the respective deductive and inductive parts, can be found in the supplementary material.

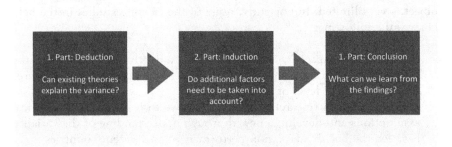

Figure 3.1 Steps of the theoretical chapters

Can existing theories explain the variance?

Since research on adaptation to climate change related health risks has only started to develop recently, not much is known yet about how and why countries differ with regards to their health adaptation to climate change.[5] Moreover, most existing studies on climate change and health originate from geography and public health research and focus more on discovering and assessing current trends rather than understanding the underlying causal relationships.

Nevertheless, in their seminal articles, Biesbroek et al. (2010), Huang et al. (2011), Lesnikowski, Ford, Berrang-Ford, Barrera, Berry, et al. (2013), Berrang-Ford et al. (2014), Austin et al. (2016), and others have identified drivers of and barriers to adaptation to climate change that will in the following be discussed and subsequently tested. At the same time, however, the drivers and barriers that were identified so far only constitute a small share of the potential factors that might explain the behavior of states with regards to climate change related health risks and the majority of these factors lacks a theoretical underpinning in the original articles. Therefore, it is necessary to take additional academic literature into account and delve more into the theoretical background of the politics of climate change adaptation to derive hypotheses that can be tested as to whether they may explain the described variances.

Exposure and vulnerability

Although climate change is a global phenomenon, it affects every country differently (Leal Filho, Azeiteiro, and Alves 2016, Smith et al. 2014). While many countries already have to deal with severe consequences of climate change, such as heatwaves, droughts, and forest fires, the effects on other countries are more indirect and lie more in the future (Smith et al. 2014). In addition to the different kinds of impacts climate change has in the world, the severity of the impact also varies greatly (Smith et al. 2014). Although some regions may even benefit from rising temperatures, since for example new resources may become available, most regions have to adapt to drastic changes that threaten the health of their populations (Corvalán, Gopalan, and Llansó 2003).

Three terms help to understand the levels of risk associated with climate change: hazard, exposure, and vulnerability (Cardona et al. 2012, 69). Hazard describes the "possible, future occurrence of natural or human-induced physical events that may have adverse effects on vulnerable and exposed elements" (Cardona et al. 2012, 69). Typical hazards are extreme weather events, such as storms or floods (Casas et al. 2016, 74). The existence of hazards alone does however not fully determine the level of risk since other factors may influence the effects of such hazards and therefore the associated risks (Cardona et al. 2012, 69).

Therefore, the second component – exposure – is essential to understanding risks. Exposure "refers to the inventory of elements in an area in which hazard events may occur" (Cardona et al. 2012, 69). As a consequence, hazards are only relevant factors to determine the risk of areas if such areas are actually exposed to the hazard. Following this reasoning, a storm, for example, may constitute a risk for a large city that is exposed to it, but it would constitute a significantly lower risk for an unpopulated area in the middle of the Sahara where no infrastructure, significant vegetation, or living beings exist. According to Cardona et al. (2012, 69), exposure constitutes a necessary, but not a significant determinant of risk. This is mostly due to the last term that needs to be introduced for a thorough understanding of risks associated with climate change: vulnerability.

According to the IPCC, vulnerability "is the degree to which a system is susceptible to, and unable to cope with, adverse effects of climate change, including climate variability and extremes" (Baede, Linden, and Verbruggen 2007, 89). Cardona et al. (2012, 70) summarize the factors that influence vulnerability of a state as "predisposition, susceptibilities, fragilities, weaknesses, deficiencies, or lack of capacities that favor adverse effects on the exposed elements." As a consequence, states may be exposed to climate change related hazards, but due to their high capacities, the risks may be lower than for states that are less exposed but have significantly lower capacities to deal with the hazards. Following the trichotomy of risk levels, the focus of this study will be on vulnerability since it is more comprehensive than hazard and exposure and can help to better understand states' risk perception and motivation to act against climate change related health risks.

In the academic literature on the drivers, facilitating factors, and barriers of adaptation to climate change, vulnerability plays an important role. While Berrang-Ford et al. (2014) and Austin et al. (2016) did not find a strong link between vulnerability (or exposure) and adaptation to climate change, Biesbroek et al. (2010) came to the conclusion that the impact of extreme weather events and other region-specific risks drive European states to develop their own adaptation plans and strategies. Massey et al. (2014) analyzed the motivations of 29 countries to strengthen adaptation measures to climate change and the respondents stated that past extreme weather events were the main reasons for them to adapt to climate change. Similarly, Tompkins et al. (2010, 632) found that, amongst other factors, the most frequently named motivation of states to strengthen their adaptation to climate change was the actual or perceived impact of climate change, as for instance through extreme weather events. To test whether a country's vulnerability to climate change related health risks really drives its measures in the field, the following hypothesis shall be tested:

> H1: The more vulnerable a country is to climate change related health risks, the more it does to adapt to climate change related health risks.[6]

Socioeconomic factors

The relationship between a country's vulnerability and its actions appears to be very straightforward: the more vulnerable a country is to climate change related health risks, the more it does to reduce its vulnerability. But what if a state is lacking the means to do so? Huang et al. (2011, 185) claim that insufficient funding for climate change adaptation is not only an issue for developing countries but also for large economic powers, such as the United States, since governments often do not assign adequate funding to adaptation efforts.

Although a significant amount of studies has emerged on the influence of socioeconomic factors on the health performance of different countries, such as the impact of social capital (Kawachi et al. 1997, Kawachi, Kennedy, and Glass 1999, Szreter and Woolcock 2004) or the population's income (Cutler, Deaton, and Lleras-Muney 2006, Keefer and Khemani 2005, Pritchett and Summers 1996, Wilkinson and Marmot 2003, Wilkinson, Campbell-Lendrum, and Bartlett 2003), research on the influence of socioeconomic factors on health adaptation to climate change is still underdeveloped. Nevertheless, some articles on how socioeconomic factors either drive or prevent adaptation to climate change in general have emerged (Lesnikowski, Ford, Berrang-Ford, Barrera, Berry, et al. 2013, Massey et al. 2014). The factors include national wealth, social expenditure, economic performance, and others (Lesnikowski, Ford, Berrang-Ford, Barrera, Berry, et al. 2013, Massey et al. 2014).

Lesnikowski, Ford, Berrang-Ford, Barrera, Berry, et al. (2013) use the GDP of states as an indicator for national wealth and come to the conclusion that wealthier and larger countries are "more likely to be engaged in a wider range of health vulnerabilities." It is however important to note that Lesnikowski, Ford, Berrang-Ford, Barrera, Berry, et al. (2013) focus on Annex 1 Parties to the UNFCCC[7] and thus have a bias towards generally more developed states. In a similar study, Massey et al. (2014) found that the motivation of countries to act against climate change is largely driven by external factors, such as the influence of international organizations, but the nominal GDP of a country has an important influence on which external factors are most relevant (Massey et al. 2014). According to Massey et al. (2014), for countries with high and medium GDP, extreme weather events and scientific research constitute the most important motivations, whereas for countries with lower GDP, agenda-setting by the EU drives their adaptation measures the most. Berrang-Ford et al. (2014) state that GDP and population size – as proxies for national wealth – can influence adaptation policies as well, although they generally focus more on institutional factors in their study, such as good governance. Moreover, Austin et al. (2016, 9) can confirm the finding of Lesnikowski, Ford, Berrang-Ford, Barrera, Berry, et al. (2013) that GDP is a driver of health adaptation. Therefore, the following hypothesis shall be tested:

> H2: The better the economic health (measured in GDP PPP[8]) and performance (measured in GDP PPP per capita) of a state is, the more it does to adapt to climate change related health risks.

Population size

The sheer size of a population can also affect how well a state adapts to climate change since, in many cases, the affected population is bigger than in other countries and at the same time more resources are available to develop and implement adaptation initiatives. Lesnikowski, Ford, Berrang-Ford, Barrera, Berry, et al. (2013, 1159) found out that countries with high population sizes tend to have more health adaptation initiatives than those with smaller populations, but the population size alone cannot explain the different performances since, as the authors' state, other factors, such as the GDP or perception of corruption, also seem to have an impact. Similarly, Berrang-Ford et al. (2014, 445) came to the conclusion that national population sizes, together with other factors, correlate with high levels of adaptation. Therefore, the following hypothesis shall be tested:

> H3: The higher the population size of a state is, the more it does to adapt to climate change related health risks.

Academic literature on adaptation to climate change discusses a variety of political factors that may either drive or hold back actions, such as the level of corruption (Berrang-Ford et al. 2014), domestic or international political pressure (Massey et al. 2014), political awareness (Massey et al. 2014), influence of international organizations (Massey et al. 2014, Biesbroek et al. 2010), and governing structures (Austin et al. 2016). Many of the existing studies, however, focus on adaptation in general and not on health adaptation in particular. Additionally, almost none of them takes a closer look at differences between political regime types. Therefore, findings from other policy areas, such as regime type effects on health policies or adaptation in general, will be considered to develop appropriate hypotheses.

Political regime types

Only a few people would deny that democratic and autocratic regimes fundamentally differ with regards to their political structures, processes, and often also their political outputs – their policies. According to Besley and Kudamatsu (2006), the key differences between democracies and autocracies are political representation, accountability structures, and the role of elections. The regularly conducted, free and fair elections of (liberal) democracies, in their view, lead to stronger mechanisms for selecting political leaders, which again positively affects their (health) policies (Besley and Kudamatsu 2006). Similar to elections, the representation of interests of the population or specific groups in the policymaking process or the level of political freedom and civil rights may influence their adaptation actions since dominant interest groups may have different goals from those of the majority of the society (Besley and Kudamatsu 2006). Therefore, comparing the performances of political regime types helps to find general trends in a very complex policy area.

Academic research on the performance of democracies and autocracies has rapidly and continuously risen in the last decades (Deacon 2009, Diamond 2002, Merkel 2013, Nelson 2007, Schmidt 2013, 2014). Following the thesis of former British Prime Minister Winston Churchill that democracy is "the worst form of government except all those other forms that have been tried from time to time," a significant amount of articles assumes that democracies have clear advantages over autocracies concerning certain normative goals, but not with regard to the performance in general (Churchill 1974, 7556, Halperin, Siegle, and Weinstein 2010, Schmidt 2010, 473). Whether democracies actually perform better than autocracies or not often depends on the specific subtype of the political regime (e.g. closed autocracy, electoral authoritarianism, defective democracy, electoral democracy, liberal democracy, etc.), the democratic tradition (how long is the state a democracy?), and the specific policy field (Berg-Schlosser and Kersting 1997, 138–139, Croissant 2004, Croissant and Kuehn 2009, Halperin et al. 2010). Singapore, for example, performs much better when it comes to corruption than most liberal democracies, although the Asian city-state can be characterized as an electoral authoritarian regime (Quah 1995, 2001).

Studies on the performance of democracies and autocracies span from the impact of regime types on sustainable development (Wurster 2011) and public and social services (Brown and Hunter 1999, Lake and Baum 2001) to a variety of other policy areas. Whilst some authors claim that autocracies have advantages over democracies when it comes to facilitating and promoting economic development (Przeworski and Limongi 1993), since they can more easily push through reforms and thus provide better conditions for economic growth, others argue that democracies strengthen the protection of property rights, which can in turn foster economic growth (Knack and Keefer 1995, Knutsen 2015, 358). Knutsen (2015, 365) argues that, in the long run, democracies perform substantially better with regards to fostering technological change, since their high levels of civil liberties improve the diffusion of new ideas, which leads to innovation.

Over the years, several studies have dealt with how political factors influence health policies across different regime types. Most of them focus on the impact of democracies and autocracies on infant, child, and maternal mortality (Álvarez-Dardet and Franco-Giraldo 2006, Franco, Álvarez-Dardet, and Ruiz 2004, Deaton, Jack, and Burtless 2004, Filmer and Pritchett 1999, Gerring, Thacker, and Alfaro 2012, Kudamatsu 2012, Navia and Zweifel 2000, 2003, Ross 2006). Several authors, such as Kudamatsu (2012) and Franco et al. (2004), claim that democratization and democratic governance have positive effects on public health. However they do not explain the underlying causal mechanisms of their observations (Franco et al. 2004, 1423).

Additionally, some studies have analyzed the role of democratic governance on more specific health determinants, such as mental health (Cohen 2009). Wise and Sainsbury (2007, 180–181), and argue that characteristics of liberal democracies, such as more participation, political rights, and civil liberties, "are associated with higher levels of happiness and/or subjective well-being." They also claim that in democratic societies more people are involved in decision making processes, which leads to better monitoring and control of government actions and more inclusive health policies, which again has positive effects on public health (Wise and Sainsbury 2007, 180–181). Franco et al. (2004, 1422) published similar findings: they evaluated health specific data from the Human Development Report and the International Monetary Fund and found that states with healthier populations also possessed higher Freedom House Ratings. Wigley and Akkoyunlu-Wigley (2011) came to the conclusion that states with longer democratic traditions show higher rates of life expectancy than young democracies. Additionally, in the same year, Wigley and Akkoyunlu-Wigley (2011, 607) observed another variance between democracies: those allowing and promoting a more proportional representation of societal preferences tend to have healthier populations.

Furthermore, democratic principles and institutions, as for instance regularly held elections, universal suffrage, civil liberties, and multi-party competition, may lead to more competition between political elites, which eventually leads to better health policies (Ruger 2005, 299–300, Wigley and Akkoyunlu-Wigley 2011, 647). Moreover, some scholars argue that autocracies have fewer incentives to improve the living conditions of their populations because of the lack of societal and political checks and balances (Wigley and Akkoyunlu-Wigley 2011, 647). Wigley and Akkoyunlu-Wigley (2011, 653) argue that free media can play a similar role in controlling government activities and can consequently lead to better health policies.

Although a number of studies on the impact of institutional factors on adaptation to climate change related health risks has been published, the influence of political factors on health adaptation to climate change has less frequently been discussed (Hesketh and Zhu 1997, 1997b, Paterson et al. 2012, Ratnapradipa 2012, Shin and Ha 2012, Wei et al. 2014). Most of the existing studies analyze the situation in specific states or regions in detailed case studies and therefore do not make any claims about general trends (Wei et al. 2014, Paterson et al. 2012). Additionally, several researchers shed light on states' reactions to specific health risks, such as infectious diseases or the consequences of extreme heatwaves, but do not provide a comprehensive analysis of the potential reasons for the variance of a great number of states on a larger number of climate change related health risks (Hess et al. 2013, Moffatt 2009, Panic and Ford 2013, White-Newsome et al. 2014).

In terms of political and institutional drivers of and barriers to general climate change adaptation, Biesbroek et al. (2010, 448) have identified that, among other factors, civil society, special events, and international institutions affect decision making in this field. The classification does, however, lack a reflection of the differences between democracies and autocracies, which could drastically influence variances between national adaptation strategies.[9] Based on the literature review, it can be concluded that substantial characteristics of liberal democracies, including higher levels of civil liberties, political rights, representation of the society in politics, and regularly held, free and fair elections, which foster political competition, have the potential to lead to more comprehensive and better health adaptation policies. Therefore, the following hypothesis shall be tested:

> H4: The more a state guarantees civil liberties, political rights, and regularly held, free and fair elections, the more it does to adapt to climate change related health risks.

Good governance

Countries do not only differ in terms of their general political systems, but also with regard to how well they are governed. This again can have significant effects on their policies. Additionally, even countries that are very similar when it comes to the overall political system and the governance structure may have fundamentally different characteristics, especially when it comes to the quality of governance. In recent years, the distinction between good and bad governance has been established in social sciences. UNESCAP (n.d., 1) defines good governance as "participatory, consensus oriented, accountable, transparent, responsive, effective and efficient, equitable and inclusive and follows the rule of law." Although some elements, such as "participatory" or "inclusive," are at least partially included in the distinction between political regime types, other aspects, such as "responsive, effective and efficient" add new functional components that are worth testing for. To operationalize good governance, numerous authors make use of Transparency International's CPI as a proxy (Berrang-Ford et al. 2014, 448). Both Lesnikowski, Ford, Berrang-Ford, Barrera, Berry, et al. (2013), and

Berrang-Ford et al. (2014, 448) found out that a relationship between governance and adaptation exists: states with good institutional governance show better adaptation initiatives than those with poor institutional governance levels. This leads to the following hypothesis:

> H5: The lower the perceived corruption of a state is, the more it does to adapt to climate change related health risks.

The role of epistemic communities and the international community

In the field of climate change and health, epistemic communities and representatives of international organizations are often significantly involved in the policy process since they support governmental agencies and other relevant stakeholders in assessing the risks and developing strategies and plans.[10] While recent articles on climate change adaptation have acknowledged that the international community and academia can play a significant role, they struggle to operationalize and measure their concrete influence on the decision making process (Massey et al. 2014, Biesbroek et al. 2010, Lesnikowski, Ford, Berrang-Ford, Barrera, Berry, et al. 2013). Through their research and policy recommendations, epistemic communities and experts at international organizations reduce some of the uncertainty related to climate change and health, which can make it easier for states to develop adaptation measures and transition to the prevention principle.

Additionally, many states strive for international reputation, which can be achieved by pursuing leadership roles in the international climate change regime. This is, for instance, the case for the RoK, which has put a lot of effort into becoming a leader in adaptation to climate change (NAP-EXPO 2017).[11] With their shared ideas and norms on climate change and health, epistemic communities and international organizations, such as the WHO, can thus lead to strengthened commitments by states to act. Nevertheless, commitments do not always directly lead to more actions since in some cases they are used to cover up lacking implemented actions.

To account for the impact of epistemic communities and international organizations on states' performance on the CHAIn and its sub-indexes, the following hypotheses shall be tested:

> H6: The stronger the influence of international organizations is on the state, the more it does to adapt to climate change related health risks.
>
> H7: The stronger the epistemic community within a state is, the more it does to adapt to climate change related health risks.

Additional factors

As the nexus between climate change and health is a rapidly evolving research topic and not much is known yet about the drivers and barriers of health adaptation on a global scale, a range of additional factors could theoretically influence how states perceive and adapt to climate change related health risks. Conversations with experts in the field have shown that generational effects may occur. States with younger populations may tend to more future-oriented policies and thus more measures to adapt to climate change related health risks since the share of voters who benefit from such policies may be higher than in countries with older populations. As a consequence, the following hypothesis shall be tested as well:

> H8: The younger the population of a state is, the more the state does to adapt to climate change related health risks.

The list of factors that were addressed here is definitely not complete and, as we continue to learn about the politics of climate change and health, more studies become necessary to investigate what drivers and holds back adaptation measures. At this point, however, the analysis has led to a comprehensive model that incorporates the above hypotheses and enriches them with a systemic approach, as the following paragraphs will demonstrate.

Building the model

The empirical parts will demonstrate later on that health adaptation to climate change constitutes a very complex challenge that contains high levels of uncertainty. Uncertainty is understood as "indeterminacy of a largely socially constructed world that lacks meaning without norms and identities" (Rathbun 2007, 533). To effectively deal with this complexity and uncertainty, states need to a) have access to what relevant actors define as information on climate change related health risks, b) recognize the need to act against such risks, c) be willing to develop health adaptation measures, and d) have sufficient resources to implement the necessary measures. The information, recognition, willingness, and resources of a state influence each other and are influenced by a variety of internal and external factors, which will in the following be discussed in detail.

Information

Since the nexus between climate change and health has only in recent years started to become a major part of climate change research, not all states have equal access to information on the specific health risks that the research community associates with climate change. Moreover, due to the high complexity and uncertainty of the topic, states vastly depend on national and international epistemic communities and international organizations, such as the WHO, to develop an understanding of the risks they are facing and what could be done to reduce those risks. Following Haas (1992, 3), control over information can largely influence the actions of states and international policy coordination. Due to the complexity and interconnectedness of environmental topics, such as climate change, Haas (1990, 1992) argues that epistemic communities have a significant influence on states' actions.

Haas (1992, 3) defines an epistemic community as a "network of professionals with recognized expertise and competence in a particular domain and an authoritative claim to policy-relevant knowledge within that domain or issue-area." Although Haas (1992, 3) claims that the members of an epistemic community may come from a variety of disciplines and backgrounds, he admits that his understanding of an epistemic community is narrower than that of Holzner and Marx (1979), who do not focus on a specific groups of scientists, and needs to "be made up of natural scientists or of professionals applying the same methodology that natural scientists do." In this respect, the definition of epistemic communities used in this thesis differs from what Haas (1992, 3) proclaims since the epistemic communities that were identified in the case studies also included social scientists and practitioners. Nevertheless, the majority of members of the epistemic communities in the field of climate change and health has a background in natural sciences or medicine.

According to Haas (1992, 3), epistemic communities share a) a set of norms and principles, which constitute the foundation of their social actions, b) causal beliefs, which are based on their professional practices and guide them towards their understanding of possible policies and desired outcomes, c) ideas of how they weigh and validate knowledge in their area of expertise, and d) practices that are directed towards certain problems to enhance human welfare. Haas (1992) understanding of epistemic communities can be applied to the climate change and health community, which shares a) norms and principles about climate change and its health impacts, b) the notion that scientific knowledge about the effects of climate change on health should be utilized to develop better health adaptation policies, c) the understanding of essential determinants of quality research on climate change and health, and d) a set of practices to enhance human welfare by protecting humanity from climate change related health risks, through strengthened research, policy proposals, and advocacy.

In terms of climate change and health, some distinct epistemic communities exist. Some of them are more research oriented and are mostly centered at universities, whereas others have a strong advocacy dimension and cooperate with representatives of international organizations to push states towards more recognition of climate change related health risks and strengthened health adaptation initiatives. The most prominent epistemic community in this regard contributes to the Lancet Countdown on Climate Change and Health and includes representatives from universities, research institutes, academic journals, and international organizations (Watts et al. 2018).

Recognition

Chapter 5 in Part II will show that the recognition levels of climate change related health risks largely differ from country to country. On the one hand, this suggests that not all information on climate change related health risks, which is available, is also received by the respective states since the flow of information depends on both the sender and the receiver (Lasswell 1948, Lanham 2003). The power of epistemic communities and international organizations, which are the main senders of information on climate change and health, varies from country to country. On the other hand, even if states receive information about the consequences of climate change on the health of their populations, they may perceive that information differently based on their reference points or they may not be willing to publicly acknowledge that information since they fear the consequences of such public recognition (Kahneman and Tversky 2013, Tversky and Kahneman 1992, Levy 1997).

The study has shown that even countries with similar vulnerability levels have different recognition levels. A growing number of academic articles has emerged recently that examines the perception of climate change risks in different countries. Jurgilevich et al. (2017) undertook a review of the climate risk and vulnerability assessments of 42 countries and came to the conclusion that risk perceptions differ largely across states, with some focusing more on current risks and having a rather short-term perspective and others taking the long-term future into account. Moreover, the components of vulnerability and risk assessments often differ, with some states including both "vulnerability and exposure dynamics" and others having a more limited assessment (Jurgilevich et al. 2017). Mayer et al. (2017, 12) came to the conclusion that the actual risks associated with climate change only play a negligible role for climate policies, but risk perceptions are largely influenced by other factors and thus lead to varying policies. They argue: "Perhaps emissions from vehicles and transportation are a more visceral, immediate reminder of ecological degradation caused by human activities than emissions from other sources" (Mayer et al. 2017, 12). The authors drew the conclusion that how states perceive environmental risks mostly depends on their "subjective considerations and ideological beliefs as opposed to actual, objective exposure to environmental risks" (Mayer et al. 2017, 14).

Their thesis supports the preliminary findings of this study, as no significant relationship between the vulnerability of states and their adaptation initiatives to climate change related health risks exists, but the case studies suggest that the risk perception of the society and key decision makers has a strong impact on their decisions. While in some states, such as Japan, climate change adaptation was for a long time perceived as something negative that threatens mitigation efforts, other states, such as the RoK or the UK, very early came to the conclusion that health adaptation to climate change is necessary and therefore fostered research and developed in the field early on.

The Large-N-Analysis (LNA) will show that the more democratic states are, the more climate change related health risks they recognize. This finding suggests that democratic institutions influence how states attribute climate change to certain health risks and how they perceive these risks. Based on the general characteristics of democracies, such as the ability of citizens to organize themselves and advocate for their interests, it can be assumed that epistemic communities are freer and stronger to transmit and receive ideas in democratic regimes. Therefore, they can spread more information about climate change and influence the recognition more decisively than in autocratic regimes (Wurster 2011, Ruger 2005, 299–300, Wigley and Akkoyunlu-Wigley 2011, 647). Additionally, media and civil society organizations soak up information from epistemic communities and put additional pressure on governments to establish a certain recognition level of climate change related health risks (Wigley and Akkoyunlu-Wigley 2011, 653).

As Figure 3.2 demonstrates, the information that states receive on climate change related health risks and their recognition of this information are in an independent relationship with each other. A semi-permeable layer of various factors lies in between them. The major factors include interest groups and the media. Most importantly, epistemic communities influence which information states receive and how it is framed, which can have a significant influence on their recognition and perception of such risks. Additionally, the kind of information that is produced and shared within a state depends on the willingness of states to provide epistemic communities with the opportunity and room to develop information and then recognize it.

Moreover, the international community largely influences which information states receive, how other factors and actors, including epistemic communities within the states, act, and whether and to what extent states are willing to attribute climate change to certain health risks (Figure 3.3). According to Finnemore (1993, 585–593), international organizations frequently act as "teachers of norms" by spreading science policies, integrating international norms into cooperation projects with the respective host countries, and interacting with national and international epistemic communities to shape the national governments' choices and set the agenda. Democracies are generally more likely to cooperate on the international level (Martin 2000, Iida 2002). As a consequence, the influence of international organizations and epistemic communities can to a great extent explain why more democratic states tend to higher recognition levels than autocratic states. Their recognition of scientific norms in the field of climate change and health can largely influence their willingness to adapt.

information

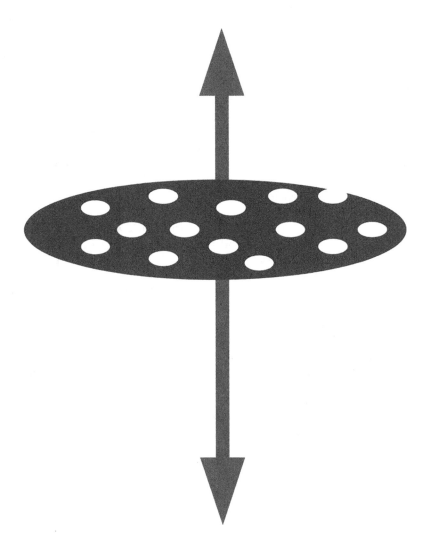

recognition

Figure 3.2 The relationship between information and recognition of climate change related health risks

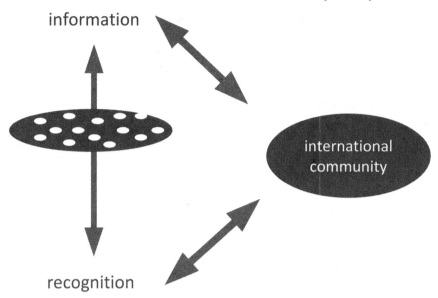

Figure 3.3 The international community and the relationship between information and recognition

Willingness

Democracies and autocracies alike are hesitant to attribute health risks to climate change in order to prevent a strong imperative to act against climate change related health risks that might constitute a great challenge to their political system since they may be under the impression that their state does not have the appropriate capacities to deal with this challenge at such a large scale. If states, for instance, directly attribute climate change to a certain number of deaths as a consequence of climate change related health risks, such as heatwaves, citizens may accuse them of not fulfilling state functions and thereby directly contributing to these mortality rates.[12] Whilst most democracies cannot neglect the general nexus between climate change and health, they are hesitant to directly link climate change to certain mortality numbers and securitize the nexus between climate change and health to an existential threat (Wæver 2011, Williams 2003). In countries with extraordinarily strong epistemic communities or influential representatives of international organizations or other states, the recognition level may cross a certain threshold and thus trigger stronger willingness to adapt.

States' willingness to invest in health adaptation to climate change very much depends on their risk perception and recognition of climate change related health risks in relation to other topics that shape the political agenda. Since the nexus between climate change and health is often perceived as a long-term policy challenge and many unknowns exist in terms of the extent and scale of specific health risks, many states hold the belief that they need to wait until further concrete information on the actual impact of climate change on health risks in their country is available and prioritize other topics on the political agenda. Other states, however, follow the precautionary principle.

Precautionary principle

The precautionary principle (in German *Vorsorgeprinzip*) is often associated with the German government's decision to incorporate the binding principle of precautionary action against potential environmental risks in its first environmental program in 1971, although the entire scale and likelihood of the risk were not fully known (Tosun 2013, 39). The key message of the precautionary principle is that under some circumstances it does not suffice to wait until complete scientific certainty exists about some risks, because the scale and speed of the risk may be so high that once it started to take place, it can no longer be averted (Tosun 2013, 39). This is particularly true for climate change since the greenhouse gas emissions from the past will persist in the atmosphere for centuries and the consequences of global warming will be felt even if no more new emissions are released (IPCC 2018, 7). Therefore, the precautionary principle suggests that those making decisions should not wait until they have scientific certainty about the respective issues, but rather act in advance in order to prepare for potential risks (Jordan and O'Riordan 1999, 23). An essential difference between the precautionary principle and other ways of dealing with the future is that the precautionary principle focuses on decision making under high levels of uncertainty (Tosun 2013, 40). It constitutes the counterpart to the prevention principle, which deals with very probable risks and therefore requires decision makers to take respective actions to prevent specific foreseeable risks (Tosun 2013, 40).

Despite some significant differences, the precautionary principle and the prevention principle share an important common trait: both are based on proaction instead of reaction to certain risks (Whiteside 2006, 146). Both concepts are centered around the belief that actions need to be taken immediately in order to prepare for the future, whereas other concepts follow the idea that it is better to wait until events unfold and then take measures that prevent similar future risks (Whiteside 2006, 61). The latter principle can be characterized as the reaction principle.

As Table 3.1 shows, the three aforementioned principles differ concerning their underlying rationales of action, the time of action, and the conditions under which decisions are being made. Whilst the precautionary principle involves action before something might potentially happen, the prevention principle has to deal with less uncertainty since the anticipated risks are very probable. In other words, similar to the precautionary principle, the prevention principle involves action before the risk unfolds, but that action should be made to prevent something that is likely to happen. Opposed to

Table 3.1 Principles of dealing with the future (own table based on Tosun 2013)

	Precautionary principle	Prevention principle	Reaction principle
Rationale	act before risks may unfold even if it is uncertain whether this will happen	prevent something that will probably happen	react after something has happened to control the damage
Time of action	before potential risk	before expected risk	after the risk materialized
Decision making under	uncertainty	high probability	certainty/high probability (informed reaction) uncertainty (uninformed reaction)

this rationale, the reaction principle means that no action takes place until something has really happened. After the initial event took place, certainty about the effects exists and action shall be taken to prevent such events from happening again (informed reaction). In some rare instances, large uncertainty continues to exist even though the event has taken place already. This might be the case when key decision makers do not have access to or do not fully understand the issue and they still react based on their limited information (uninformed reaction).[13]

The role of epistemic communities and the international community

In the field of climate change and health, epistemic communities and representatives of international organizations are often significantly involved in the policy process since they support governmental agencies and other relevant stakeholders in assessing the risks and developing strategies and plans.[14] As a consequence, they not only influence the recognition levels of states, but also whether they lean towards the precautionary principle, the prevention principle, or the reaction principle. Through their research and policy recommendations, they reduce some of the uncertainty related to climate change and health, which can make it easier for states to develop adaptation measures and transition to the prevention principle.

States' willingness to adapt to climate change related health risks cannot be simply explained by different exposure or vulnerability levels because, ultimately, the vulnerability of states is a social construction (it is a perceived vulnerability). As the case studies have shown, the understanding of health and climate change largely differs across countries and so does their perception of climate change related health risks.[15] To use the words of Alexander Wendt (1992), vulnerability often is "what states make of it." What they make of it is often framed by epistemic communities and their identity is built up over a long time through language, formative events and other identity-shaping factors. Due to their general openness towards epistemic communities, civil society organizations, and international organizations, democracies tend to be more receptive towards new ideas, which is reflected in their awareness of climate change related health risks and their higher willingness to adapt than autocracies (Nelkin 1979).

Additionally, many states strive for international reputation, which can be achieved by pursuing leadership roles in the international climate change regime. This is for instance the case for the RoK, which has put a lot of effort into becoming a leader in adaptation to climate change (NAP-EXPO 2017).[16] With their shared ideas and norms on climate change and health, epistemic communities and international organizations, such as the WHO, can thus lead to strengthened commitments by states to act. Nevertheless, commitments do not always directly lead to more actions since in some cases they are used to cover up lacking implemented actions.

To summarize, the willingness of states to adapt to climate change related health risks depends on a number of factors, including how they receive and perceive information about climate change and how they attribute certain health risks to climate change. As a consequence, for information to lead to adaptation-willingness, it needs to go through various semi-permeable layers. How the information is filtered depends on various internal and external factors, such as epistemic communities, civil society, and international organizations.[17] How the factors come into play again depends on the political regime type. However, even if states are willing to adapt, they often do

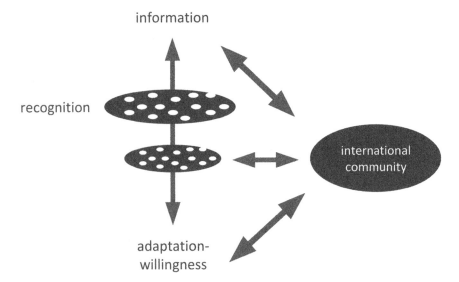

Figure 3.4 The interdependencies between information, recognition, and adaptation-willingness

not have adequate resources to put their strategies and plans into tangible adaptation-level measures.

Resources and action

How many and which adaptation-level measures states implement to reduce climate change related health risks largely depends on their resources to do so (Berrang-Ford et al. 2014, Lesnikowski, Ford, Berrang-Ford, Barrera, Berry, et al. 2013). Although the willingness to allocate appropriate resources to health adaptation to a great extent depends on the risk perception and the value states associate with reducing climate change related health risks compared with other policy areas, clear differences concerning the available means to adapt exist. This includes financial and technical resources as well as expertise and experience within the responsible institutions.[18] Therefore, states with a high GDP (PPP) tend towards having more adaptation-level measures.[19]

However, the international community often steps in and provides the resources that states are lacking. International organizations, such as the UNDP, the WHO, or the World Bank, and some high-performing states, such as the UK or Germany, support other states in developing their strategies and plans, finance research projects, and oversee the implementation of concrete health adaptation projects.[20] Accordingly, even states with comparatively low economic performances, such as Sri Lanka, the Solomon Islands, or Cyprus, can rank high on the CHAIn if they receive adequate support from the international community.

This argument shows that, together with epistemic communities, international organizations and other states can have a significant influence on the individual levels that ultimately determine how states adapt to climate change related health risks. Despite varying risks and risk perceptions, political systems and institutions, and other determinants of health adaptation to climate change, states can have similar health adaptation performances if the international community provides resources that states themselves do not possess. Nevertheless, the support of international organizations needs to be taken with a grain of salt: as the case studies have shown, the success and sustainability of international projects on health adaptation to climate change often depend on how the projects are designed and what happens with the initiated processes after the official part of the projects is finished.[21] In some countries, international communities may bring in external expertise through consultants and other international stakeholders and develop excellent strategies and plans, which local populations, however, cannot implement due to lacking adequate resources.[22] According to national experts, it requires a transformation from project ownership to policy ownership in the host countries as well as adequate funding mechanisms and a multi-sectoral approach to ensure the sustainability and effectiveness of health adaptation measures.[23]

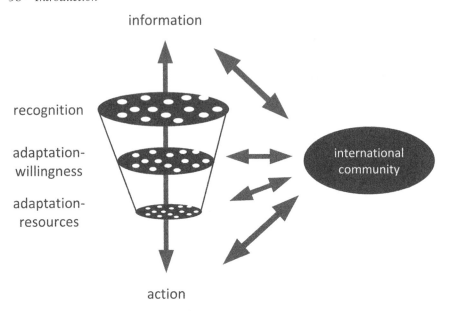

information

recognition

adaptation-
willingness

adaptation-
resources

international
community

action

Figure 3.5 Sieve-model of health adaptation to climate change

The policy cycle

The analysis has shown that epistemic communities, international organizations, or even states may largely influence the information on climate change related health risks that states receive and perceive, as well as their willingness to adapt and the availability of resources to do so. Moreover, the research has shown that the more democratic states are, the more susceptible they are for the influence of epistemic and international communities, which can have a strong influence on their recognition levels and how they communicate about climate change and health. But what does this mean for the actual policy process?

Based on the works of Easton (1965), Lasswell (1956), and others, May and Wildavsky (1979), Jenkins (1978), Brewer (1983), Anderson (2014, 1976), and Kingdon and Thurber (1984) developed the so-called "policy cycle" to contribute to a better understanding of how policies come into existence. The first stage of the policy cycle includes the problem definition, which means that the problem itself has been identified and the need for political action to solve the problem has been expressed (Jann and Wegrich 2006, 43). After the problem has been defined, it is added to the agenda, which can be described as the government's list of current priorities (Jann and Wegrich 2006, 43). The agenda setting is followed by formulating, implementing, and evaluating policies (Jann and Wegrich 2006, 43–52). If the resulting solution is deemed effective, the policy process is terminated; if not, the problem definition starts again (Jann and Wegrich 2006, 54–55). For

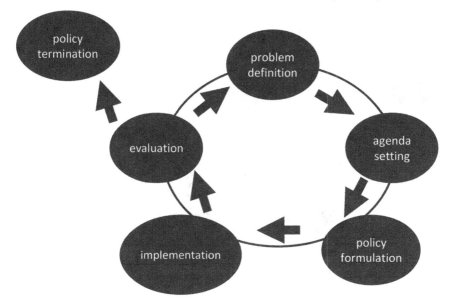

Figure 3.6 The policy cycle

the vast majority of policy problems, the process is never really terminated but rather paused until additional problems that are connected to the problem are discovered at a later point in time. Moreover, the policy cycle almost never occurs in its ideal form, but the different steps often happen simultaneously, in reverse order, or repeatedly. Nevertheless, the cycle helps to better understand the different stages ideas need to pass in order to eventually become policies.

Transferred to the complex and uncertain policy problem of health adaptation to climate change, the funnel model of information, recognition, adaptation-willingness, adaptation-resources, and adaptation action can be applied to the policy cycle (see Figure 3.7).

What Jann and Wegrich (2006, 43) describe as problem definition, can be characterized as the recognition of climate change related health risks, since the authors specify that this step involves political actors acknowledging that the problem exists and needs to be solved. Recognition itself does, however, not lead to direct action since it often requires additional internal or external pressure on the decision makers to put the topic on the agenda and formulate respective policies. Due to the high complexity and uncertainty that the nexus between climate change and health entails, the epistemic community, in cooperation with the international community, can create this pressure by publishing research findings and connecting empirical findings to extreme weather events. As a consequence, epistemic and international communities can largely influence whether extreme weather events are attributed to climate change and how states perceive the risks associated with those events.

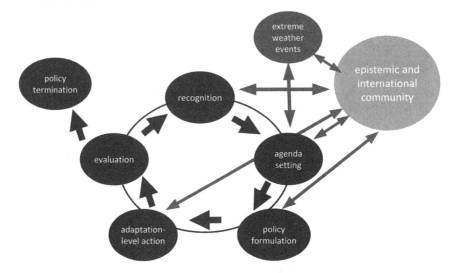

Figure 3.7 The policy cycle for health adaptation to climate change

Moreover, through international conferences and agreements, the international community pushes the topic on the agenda from time to time, reminding states of the necessity to formulate adequate policies. Additionally, the epistemic community, international organizations, and other states are often directly involved in policy formulation on the domestic level and thus have a significant impact on adaptation. Since health adaptation to climate change requires a high amount of expertise and resources, in developing countries the adaptation-level action is also often conducted in cooperation with international partners, such as the World Bank.

Overall, it can be assumed that how states perceive climate change related health risks and how they respond to such risks largely depends on how active the epistemic communities and the international community in the respective countries are. In democracies, epistemic communities and the international community have more freedom to unfold and influence the political agenda, which can explain the higher recognition levels in countries with more political liberties and civil rights. Additionally, countries with stronger economies have more means to deal with complex long-term policy challenges and therefore achieve higher adaptation-level scores. However, at times the international community steps in to develop adaptation-level measures with states and provide resources that states with lower performing economies do not have.

Notes

1 The Tracking Adaptation to Climate Change Consortium (TRAC[3]) seeks to "facilitate new collaborations that address conceptual, methodological, and practical challenges associated with tracking progress on adaptation" (TRAC3 n.d.).

2 Groundwork-level initiatives build adaptive capacity, prepare for adaptation actions, or enable adaptation actions. They include assessments, research projects, strategies, plans, recommendations, and other measures to enhance states' preparedness for climate change related health risks, but do not include concrete implemented actions, such as infrastructure measures. For further information see p. 87.

3 For detailed information on the variance between the different performances see the results of the CHAIn in Part II.

4 To learn more about mixing methods and the research design of this particular study, consider the supplementary material of this book.

5 So far only a number of case studies and comparisons of smaller groups, such as the Annex 1 countries, have been published (Berrang-Ford et al. 2014, Lesnikowski, Ford, Berrang-Ford, Barrera, and Heymann 2013, Biagini et al. 2014, Biesbroek et al. 2010, Ford et al. 2013, Juhola and Westerhoff 2011, Massey et al. 2014, Tompkins and Eakin 2012).

6 The performance of states is measured by their CHAIn score in Part II.

7 According to the UNFCCC, "Annex I Parties include the industrialized countries that were members of the OECD (Organisation for Economic Co-operation and Development) in 1992, plus countries with economies in transition (the EIT Parties), including the Russian Federation, the Baltic States, and several Central and Eastern European States" (UNFCCC n.d.)

8 While GDP is certainly not a perfect index to measure the economic health of a country, it constitutes one of the most widely used and prominently discussed factors that correlates with state action. In future models, alternatives, such as the National Index of Welfare, can be tested as well: http://www.fest-nwi.de/en/index.

9 Nevertheless, the factors identified by Biesbroek et al. (2010) and other authors contain valuable and valid potential explaining factors.

10 These ideal types are rarely found in reality since it is not always possible to clearly distinguish the degrees of certainty and to categorize decision makers' level of information. Consequently, real-life decisions are often a mix of various principles. Nonetheless, the overall principle still shines through most policies if the key decisions of several states are compared. This is particularly true for challenges like climate change, where generally high levels of uncertainty exist.

11 For more information on the RoK's pursuit of reputation through a leadership position in adaptation to climate change see Part III.

12 This argument is in some states already utilized by civil society organizations, such as Extinction Rebellion in the UK.

13 For more information see Part III.

14 These ideal types are rarely found in reality since it is not always possible to clearly distinguish the degrees of certainty and to categorize decision makers' level of information. Consequently, real-life decisions are often a mix of various principles. Nonetheless, the overall principle still shines through most policies if the key decisions of several states are compared. This is particularly true for challenges like climate change, where generally high levels of uncertainty exist.

15 See the case studies in Part III.

16 For more information on the RoK's pursuit of reputation through a leadership position in adaptation to climate change see Part III.

17 From an International Relations perspective, the sieve model is built on a cross between republican and ideational liberalism based on Moravcsik (1997) as well as social constructivism in the tradition of Wendt (1992).

18 For more information see Chapter 12 on Sri Lanka.

19 Since Lipset (1959), various scholars have analyzed and discussed the relationship between political regime types and economic development. While Przeworski

and Limongi (1993) could not confirm that democratization leads to economic growth, Diamond (1992) argues based on Sklar (1987) that the characteristics of democracies suggest that they have a positive impact on economic development since political participation, civil liberties, and pluralistic societies were conducive to economic development.

20 For further information see Chapter 12 on Sri Lanka. Moreover, numerous NCs and national adaptation strategies have been funded or co-authored by international organizations, such as the UNDP, or other states, for instance through the German Federal Development Agency (GIZ).

21 For more information see Chapter 12 on Sri Lanka.

22 For more information see Chapter 12 on Sri Lanka.

23 For more information see Chapter 12 on Sri Lanka.

References

Álvarez-Dardet, Carlos, and Álvaro Franco-Giraldo. 2006. "Democratisation and health after the fall of the Wall." *Journal of Epidemiological and Community Health* 60 (8): 669–671.

Anderson, James E. 1976. *Cases in Public Policy-making*. New York: Praeger.

Anderson, James E. 2014. *Public Policymaking*. Cengage Learning.

Araos, Malcolm, Stephanie E. Austin, Lea Berrang-Ford, and James D. Ford. 2016. "Public health adaptation to climate change in large cities: a global baseline." *International Journal of Health Services* 46 (1): 53–78.

Araos, Malcolm, Lea Berrang-Ford, James D. Ford, Stephanie E. Austin, Robbert Biesbroek, and Alexandra Lesnikowski. 2016. "Climate change adaptation planning in large cities: a systematic global assessment." *Environmental Science and Policy* 1813: 1–8.

Austin, Stephanie E., Robbert Biesbroek, Lea Berrang-Ford, James D. Ford, Stephen Parker, and Manon D. Fleury. 2016. "Public health adaptation to climate change in OECD countries." *International Journal of Environmental Research and Public Health* 13 (889): 1–20. doi:10.3390/ijerph13090889.

Austin, Stephanie E., James D. Ford, Lea Berrang-Ford, Robbert Biesbroek, and Nancy A. Ross. 2019. "Enabling local public health adaptation to climate change." *Social Science & Medicine* 220: 236–244. doi:10.1016/j.socscimed.2018.11.002.

Austin, Stephanie E., James D. Ford, Lea Berrang-Ford, Robbert Biesbroek, Jale Tosun, and Nancy A. Ross. 2018. "Intergovernmental relations for public health adaptation to climate change in the federalist states of Canada and Germany." *Global Environmental Change* 52: 226–237. doi:10.1016/j.gloenvcha.2018.07.010.

Baede, Alfons, Paul van der Linden, and Aviel Verbruggen. 2007. "Annex II – glossary." In *Climate Change 2007: Synthesis Report. Contribution of Working Groups I, II and III to the Fourth Assessment Report of the Intergovernmental Panel on Climate Change*, edited by IPCC. Geneva: IPCC.

Berg-Schlosser, Dirk, and Norbert Kersting. 1997. "Warum weltweit Demokratisierung? – Zur Leistungsbilanz demokratischer und autoritärer Regime." In *Demokratieexport in die Länder des Südens?*, edited by R. Hanisch, 93–144. Hamburg.

Berrang-Ford, Lea, James D. Ford, Alexandra Lesnikowski, Carolyn Poutiainen, Magda Barrera, and S. Jody Heymann. 2014. "What drives national adaptation? A global assessment." *Climatic Change* 124: 441–450.

Berrang-Ford, Lea, James D. Ford, and Jaclyn Paterson. 2011. "Are we adapting to climate change?" *Global Environmental Change* 21 (1): 25–33. doi:10.1016/j.gloenvcha.2010.09.012.

Berry, Peter, M. Paddy Enright, Joy Shumake-Guillemot, Elena Villalobos Prats, and Diarmid Campbell-Lendrum. 2018. "Assessing health vulnerabilities and adaptation to climate change: a review of international progress." *International Journal of Environmental Research and Public Health* 15 (12). doi:10.3390/ijerph15122626.

Besley, Timothy, and Masayuki Kudamatsu. 2006. "Health and democracy." *The American Economic Review* 96 (2): 313–318.

Biagini, Bonizella, Rosina Bierbaum, Missy Stults, Saliha Dobardzic, and Shannon M. McNeeley. 2014. "A typology of adaptation actions: a global look at climate adaptation actions financed through the Global Environment Facility." *Global Environmental Change* 25: 97–108. doi:10.1016/j.gloenvcha.2014.01.003.

Biesbroek, G. Robbert, Rob J. Swart, Timothy R. Carter, Caroline Cowan, Thomas Henrichs, Hanna Mela, Michael D. Morecroft, and Daniela Rey. 2010. "Europe adapts to climate change: comparing National Adaptation Strategies." *Global Environmental Change* 20 (3): 440–450. doi:10.1016/j.gloenvcha.2010.03.005.

Brewer, Garry D. 1983. "Peter deLeon." *The Foundations of Policy Analysis*. Homewood, IL: Dorsey Press.

Brown, David S., and Wendy Hunter. 1999. "Democracy and social spending in Latin America, 1980–92." *The American Political Science Review* 93 (4): 779–790.

Cardona, O. D., M. K. van Aalst, J. Birkmann, M. Fordham, G. McGregor, R. Perez, R. S. Pulwarty, E. L. F. Schipper, and B. T. Sinh. 2012. "Determinants of risk: exposure and vulnerability." In *A Special Report of Working Groups I and II of the Intergovernmental Panel on Climate Change (IPCC)*, edited by C. B. Field, V. Barros, T. F. Stocker, D. Qin, D. J. Dokken, K. L. Ebi, M. D. Mastrandrea, K. J. Mach, G.-K. Plattner, S. K. Allen, M. Tignor and P. M. Midgley, 65–108. Cambridge/New York: Cambridge University Press.

Casas, André Luís Foroni, Gabriella Mendes Dias Santos, Natalia Bíscaro Chiocheti, and Mônica de Andrade. 2016. "Effects of temperature variation on the human cardiovascular system: a systematic review." In *Climate Change and Health: Improving Resilience and Reducing Risks*, edited by Walter Leal Filho, Ulisses de Miranda Azeiteiro and Fátima Alves, 73–87. Cham/Heidelberg: Springer.

Churchill, Winston. 1974. *Winston S. Churchill: His Complete Speeches, 1897–1963. Vol. VII 1943–1949*. New York/London: Chelsea House.

Cohen, Mark. 2009. "Democracy and mental health – a psychoanalytic contribution." *Journal of Public Mental Health* 7 (3): 8–14.

Corvalán, C. F., H. N. B. Gopalan, and P. Llansó. 2003. "Conclusions and recommendations for action." In *Climate Change and Human Health – Risks and Responses*, edited by A. J. McMichael, D. H. Campbell-Lendrum, C. F. Corvalán, K. L. Ebi, A. K. Githeko, J. D. Scheraga and A. Woodward, 267–283. Geneva: World Health Organization.

Croissant, Aurel. 2004. "From transition to defective democracy – mapping Asian democratization." *Democratization* 11 (5): 156–179.

Croissant, Aurel, and David Kuehn. 2009. "Patterns of civilian control of the military in East Asia's new democracies." *Journal of East Asian Studies* 9 (2): 187–207.

Cutler, David, Angus Deaton, and Adriana Lleras-Muney. 2006. "The determinants of mortality." *The Journal of Economic Perspectives* 20 (3): 97–120. doi:10.1257/jep.20.3.97.

Deacon, Robert T. 2009. "Public good provision under dictatorship and democracy." *Public Choice* 139 (1): 241–262.

Deaton, Angus, William Jack, and Gary Burtless. 2004. "Health in an age of globalization [with comments and discussion]." *Brookings Trade Forum* 2004 (1): 83–130. doi:10.1353/btf.2005.0004.

Diamond, Larry. 1992. "Economic development and democracy reconsidered." *American Behavioral Scientist* 35 (4–5): 450–499.

Diamond, Larry. 2002. "Thinking about hybrid regimes." *Journal of Democracy* 13 (2): 21–35.

Dupuis, J., and G. R. Biesbroek. 2013. "Comparing apples and oranges: the dependent variable problem in comparing and evaluating climate change adaptation policies." *Global Environmental Change: Human and Policy Dimensions* 23 (6): 1476–1487. doi:10.1016/j.gloenvcha.2013.07.022.

Easton, D. 1965. *A Systems Analysis of Political Life*. New York: Wiley.

Ebi, Kristie, and Mariam Otmani del Barrio. 2017. "Lessons learned on health adaptation to climate variability and change: experiences across low- and middle-income countries." *Environmental Health Perspectives* 125 (6): 065001. doi:10.1289/EHP405.

Filmer, Dean, and Lant Pritchett. 1999. "The impact of public spending on health: does money matter?" *Social Science & Medicine* 49: 1309–1323.

Finnemore, Martha. 1993. "International organizations as teachers of norms: the United Nations Educational, Scientific, and Cultural Organization and science policy." *International Organization* 47 (4): 565–597.

Ford, James D., and Lea Berrang-Ford. 2016. "The 4Cs of adaptation tracking: consistency, comparability, comprehensiveness, coherency." *Mitigation and Adaptation Strategies for Global Change* 21: 839–859.

Ford, James D., Lea Berrang-Ford, Alex Lesnikowski, Magda Barrera, and S. Jody Heymann. 2013. "How to track adaptation to climate change: a typology of approaches for national-level application." *Ecology and Society* 18 (3). doi:10.5751/ES-05732-180340.

Franco, Álvaro, Carlos Álvarez-Dardet, and Maria Teresa Ruiz. 2004. "Effect of democracy on health: ecological study." *BMJ* 329 (1421–1423).

Gerring, John, Strom C. Thacker, and Rodrigo Alfaro. 2012. "Democracy and human development." *The Journal of Politics* 74 (1): 1–17. doi:10.1017/S0022381611001113.

Haas, Peter M. 1990. "Obtaining International environmental protection through epistemic consensus." *Millennium* 19 (3): 347–363. doi:10.1177/03058298900190030401.

Haas, Peter M. 1992. "Introduction: epistemic communities and international policy coordination." *International Organization* 46 (1): 1–35. doi:10.1017/S0020818300001442.

Halperin, Morton, Joe Siegle, and Michael M. Weinstein. 2010. *The Democracy Advantage: How Democracies Promote Prosperity and Peace*. Vol. 2. New York: Routledge.

Hesketh, Therese, and Wei Xing Zhu. 1997. "Health in China: from Mao to market reform." *British Medical Journal* 314 (7093): 1543–1545. doi:10.1136/bmj.329.7480.1427.

Hess, Jeremy J., Paul J. Schramm, and George Luber. 2014. "Public health and climate change adaptation at the federal level: one agency's response to Executive

Order 13514." *American Journal of Public Health* 104 (3): e22–e30. doi:10.2105/AJPH.2013.301796.

Hess, Jermey J., Gino Marinucci, Paul J. Schramm, Arie Manangan, and George Luber. 2013. "Management of climate change adaptation at the United States Centers of Disease Control and Prevention." In *Global Climate Change and Public Health*, edited by Kent E. Pinkerton and William N. Rom, 341–360. New York: Springer.

Holzner, Burkart, and John Marx. 1979. *Knowledge Application: The Knowledge System in Society*. Boston: Allyn & Bacon.

Huang, Cunrui, Pavla Vaneckova, Xiaoming Wang, Gerry FitzGerald, Yuming Guo, and Shilu Tong. 2011. "Constraints and barriers to public health adaptation to climate change: a review of the literature." *American Journal of Preventive Medicine* 40 (2): 183–190. doi:10.1016/j.amepre.2010.10.025.

Iida, Keisuke. 2002. "*Democratic Commitments: Legislatures and International Cooperation*. Lisa L. Martin. Princeton, NJ: Princeton University Press, 2000, 225 pp." Book review. *International Relations of the Asia-Pacific* 2 (2): 267–269. doi:10.1093/irap/2.2.267.

IPCC. 2018. "Summary for policymakers." In *Global warming of 1.5°C. An IPCC Special Report on the impacts of global warming of 1.5°C above pre-industrial levels and related global greenhouse gas emission pathways, in the context of strengthening the global response to the threat of climate change, sustainable development, and efforts to eradicate poverty*, edited by V. Masson-Delmotte, P. Zhai, H. O. Pörtner, D. Roberts, J. Skea, P. R. Shukla, A. Pirani, W. Moufouma-Okia, C. Péan, R. Pidcock, S. Connors, J. B. R. Matthews, Y. Chen, X. Zhou, M. I. Gomis, E. Lonnoy, T. Maycock, M. Tignor and T. Waterfield. Geneva.

Jann, Werner, and Kai Wegrich. 2006. "Theories of the policy cycle." In *Handbook of Public Policy Analysis: Theory, Politics, and Methods*, edited by Frank Fischer, Gerald J. Miller and Mara S. Sidney, 43–62. Boca Raton: CRC Press.

Jenkins, William Ieuan. 1978. *Policy Analysis: A Political and Organisational Perspective*. London: M. Robertson.

Jordan, Andrew, and Timothy O'Riordan. 1999. "The precautionary principle in contemporary environmental policy and politics." In *Protecting Public Health and the Environment – Implementing the Precautionary Principle*, edited by Carolyn Raffensperger and Joel A. Tickner, 15–35. Washington DC: Covelo.

Juhola, Sirkku, and Lisa Westerhoff. 2011. "Challenges of adaptation to climate change across multiple scales: a case study of network governance in two European countries." *Environmental Science and Policy* 14 (3): 239–247. doi:10.1016/j.envsci.2010.12.006.

Jurgilevich, Alexandra, Aleksi Räsänen, Fanny Groundstroem, and Sirkku Juhola. 2017. "A systematic review of dynamics in climate risk and vulnerability assessments." *Environmental Research Letters* 12 (1): 013002. doi:10.1088/1748-9326/aa5508.

Kahneman, Daniel, and Amos Tversky. 2013. "Prospect theory: an analysis of decision under risk." In *Handbook of the Fundamentals of Financial Decision Making: Part I*, edited by L. C. MacLean and W. T. Ziemba, 127. London: World Scientific.

Kawachi, I., B. P. Kennedy, and R. Glass. 1999. "Social capital and self-rated health: a contextual analysis." *American Journal of Public Health* 89 (8): 1187–1193. doi:10.2105/AJPH.89.8.1187.

Kawachi, I., B. P. Kennedy, K. Lochner, and D. Prothrow-Stith. 1997. "Social capital, income inequality, and mortality." *American Journal of Public Health* 87 (9): 1491–1498. doi:10.2105/AJPH.87.9.1491.

Keefer, Philip, and Stuti Khemani. 2005. "Democracy, public expenditures, and the poor: understanding political incentives for providing public services." *The World Bank Research Observer* 20 (1): 1–27.

Kim, Rokho, Anthony Costello, and Diarmid Campbell-Lendrum. 2015. "Climate change and health in Pacific island states." *Bull World Health Organ* 93: 819. doi:10.2471/BLT.15.166199.

Kingdon, John W., and James A.Thurber. 1984. *Agendas, Alternatives, and Public Policies*. Vol. 45. Boston: Little, Brown and Company.

Knack, Steven, and Philip Keefer. 1995. "Institutions and economic performance: cross-country tests using alternative measures." *Economics and Politics* 7: 207–227.

Knutsen, Carl Henrik. 2015. "Why democracies outgrow autocracies in the long run: civil liberties, information flows and technological change." *Kykklos* 68 (3): 357–384.

Kudamatsu, Masayuki. 2012. "Has democratization reduced infant mortality in Sub-Saharan Africa? Evidence from micro data." *Journal of the European Economic Association* 10 (6): 1294–1317.

Lake, David, and Matthew Baum. 2001. "The invisible hand of democracy: political control and the provision of public services." *Comparative Political Studies* 34: 587–621.

Lanham, Richard. 2003. *Analyzing Prose*. London: A&C Black.

Lasswell, Harold D. 1948. "The structure and function of communication in society." *The Communication of Ideas* 37: 215–228.

Lasswell, Harold Dwight. 1956. *The Decision Process: Seven Categories of Functional Analysis*. Bureau of Governmental Research, College of Business and Public Administration, University of Maryland.

Leal Filho, Walter, Ulisses de Miranda Azeiteiro, and Fátima Alves. 2016. "Climate change and health: an overview of the issues and needs." In *Climate Change and Health: Improving Resilience and Reducing Risks*, edited by Walter Leal Filho, Ulisses de Miranda Azeiteiro and Fátima Alves, 1–11. Cham/Heidelberg: Springer.

Lesnikowski, A. C., J. D. Ford, L. Berrang-Ford, J. A. Paterson, M. Barrera, and S. J. Heymann. 2011. "Adapting to health impacts of climate change: a study of UNFCCC Annex I parties." *Environmental Research Letters* 6: 1–9. doi:10.1088/1748-9326/6/4/044009.

Lesnikowski, A. C., J. D. Ford, L. Berrang-Ford, M. Barrera, P. Berry, J. Henderson, and S. J. Heymann. 2013. "National-level factors affecting planned, public adaptation to health impacts of climate change." *Global Environmental Change* 23 (5): 1153–1163. doi:10.1016/j.gloenvcha.2013.04.008.

Lesnikowski, Alexandra C., James D. Ford, Lea Berrang-Ford, Magda Barrera, and Jody Heymann. 2013. "How are we adapting to climate change? A global assessment." *Mitigation and Adaptation Strategies for Global Change* 1–17. doi:10.1007/s11027-013-9491-x.

Lesnikowski, Alexandra, James Ford, Robbert Biesbroek, Lea Berrang-Ford, and S. Jody Heymann. 2015. "National-level progress on adaptation." *Nature Climate Change* 6: 261–265. doi:10.1038/NCLIMATE2863.

Levy, Jack S. 1997. "Prospect theory, rational choice, and international relations." *International Studies Quarterly* 41 (1): 87–112.

Lipset, Seymour Martin. 1959. "Some social requisites of democracy: economic development and political legitimacy." *American Political Science Review* 53 (1): 69–105.

Martin, Lisa L. 2000. *Democratic Commitments. Legislatures and International Cooperation.* Princeton University Press.

Massey, E., D. Huitema, A. Jordan, and G. R. Biesbroek. 2014. "Climate policy innovation: the adoption and diffusion of adaptation policies across Europe." *Global Environmental Change: Human and Policy Dimensions* 29: 434–443. doi:10.1016/j.gloenvcha.2014.09.002.

May, Judith V., and Aaron B. Wildavsky. 1979. *The Policy Cycle.* Vol. 5. London: SAGE Publications.

Mayer, Adam, Tara O'Connor Shelley, Ted Chiricos, and Marc Gertz. 2017. "Environmental risk exposure, risk perception, political ideology and support for climate policy." *Sociological Focus* 50 (4): 309–328. doi:10.1080/00380237.2017. 1312855.

McIver, Lachlan, Rokho Kim, Alistair Woodward, Simon Hales, Jeffery Spickett, Dianne Katscherian, Masahiro Hashizume, Yasushi Honda, Ho Kim, Steven Iddings, Jyotishma Naicker, Hilary Bambrick, J. McMichael Anthony, and L. Ebi Kristie. 2016. "Health impacts of climate change in pacific island countries: a regional assessment of vulnerabilities and adaptation priorities." *Environmental Health Perspectives* 124 (11): 1707–1714. doi:10.1289/ehp.1509756.

Merkel, Wolfgang. 2013. "Vergleich politischer Systeme: Demokratien und Auto-kratien." In *Studienbuch Politikwissenschaft*, edited by Manfred G. Schmidt, Frieder Wolf and Stefan Wurster, 207–236. Wiesbaden: Verlag Springer.

Moffatt, Hannah. 2009. *Communities Adapting to Climate Change: Emerging Public Health Strategies.* PhD thesis, Simon Fraser University.

Moravcsik, Andrew. 1997. "Taking preferences seriously: a liberal theory of international politics." *International Organization* 51 (4): 513–553.

NAP-EXPO. 2017. *"Regional NAP expo."* Accessed 2 June 2018. http://napexpo.org/asia/.

Navia, Patricio, and Thomas D. Zweifel. 2000. "Democracy, dictatorship, and infant mortality." *Journal of Democracy* 11 (2): 99–114.

Navia, Patricio, and Thomas D. Zweifel. 2003. "Democracy, dictatorship, and infant mortality revisited." *Journal of Democracy* 14 (3): 90–103.

Nelkin, Dorothy. 1979. "Scientific knowledge, public policy, and democracy: a review essay." *Knowledge* 1 (1): 106–122. doi:10.1177/107554707900100106.

Nelson, Joan M. 2007. "Elections, democracy, and social services." *Studies in Comparative International Development* 41 (4): 79–97.

Panic, Mirna, and James Ford. 2013. "A review of national-level adaptation planning with regards to the risks posed by climate change on infectious diseases in 14 OECD nations." *International Journal of Environmental Research and Public Health* 10: 7083–7109.

Paterson, Jaclyn A., James D. Ford, Lea Berrang Ford, Alexandra Lesnikowski, Peter Berry, Jim Henderson, and Jody Heymann. 2012. "Adaptation to climate change in the Ontario public health sector." *BMC Public Health* 12 (1): 452–452. doi:10.1186/1471-2458-12-452.

Pritchett, Lant, and Lawrence Summers. 1996. "Wealthier is healthier." *Journal of Human Resources* 31: 841–868.

Przeworski, Adam, and Fernando Limongi. 1993. "Political regimes and economic growth." *Journal of Economic Perspectives* 7 (3): 51–69.

Quah, John. 1995. "Controlling corruption in city-states: a comparative study of Hong Kong and Singapore." *Crime, Law & Social Change* 22: 391–414.

Quah, John. 2001. "Combatting corruption in Singapore: what can be learned?" *Journal of Contingencies and Crisis Management* 9 (1): 29–35.

Rakotoarison, Norohasina, Nirivololona Raholijao, M. Lalao Razafindramavo, A. Zo Rakotomavo, Alain Rakotoarisoa, S. Joy Guillemot, J. Zazaravaka Randriamialisoa, Victor Mafilaza, A. Voahanginirina Ramiandrisoa, Rhino Rajaonarivony, Solonomenjanahary Andrianjafinirina, Venance Tata, C. Manuela Vololoniaina, Fanjasoa Rakotomanana, and M. Volahanta Raminosoa. 2018. "Assessment of risk, vulnerability and adaptation to climate change by the health sector in Madagascar." *International Journal of Environmental Research and Public Health* 15 (12). doi:10.3390/ijerph15122643.

Rathbun, Brian C. 2007. "Uncertain about uncertainty: understanding the multiple meanings of a crucial concept in international relations theory." *International Studies Quarterly* 51 (3): 533–557.

Ratnapradipa, Dhitinut. 2012. "Vulnerability to potential impacts of climate change: adaptation and risk communication strategies for environmental health practitioners in the United Kingdom." *Journal of Environmental Health* 76 (8): 28–33.

Ross, Michael. 2006. "Is democracy good for the poor?" *American Journal of Political Science* 50 (4): 860–874.

Ruger, Jennifer. 2005. "Democracy and health." *Q J Med* 98: 299–304.

Schmidt, Manfred G. 2010. *Demokratietheorien – Eine Einführung*. Bonn.

Schmidt, Manfred G. 2013. "Staatstätigkeit in Demokratien und Autokratien." In *Autokratien im Vergleich*, edited by Steffen Kailitz and Patrick Köllner, 418–437. Baden-Baden.

Schmidt, Manfred G. 2014. "Public policy in autocracies and democracies." In *Comparing Autocracies in the Early Twenty-first Century*, edited by Aurel Croissant, SteffenKailitz, PatrickKöllner and Stefan Wurster, 39–56. London/New York: Routledge.

Schnitter, Rebekka, Marielle Verret, Peter Berry, Tanya Chung Tiam Fook, Simon Hales, Aparna Lal, and Sally Edwards. 2018. "An assessment of climate change and health vulnerability and adaptation in Dominica." *International Journal of Environmental Research and Public Health* 16 (1). doi:10.3390/ijerph16010070.

Shin, Yong Seung, and Jongsik Ha. 2012. "Policy directions addressing the public health impact of climate change in South Korea: climate-change health adaptation and mitigation program." *Environmental Health and Toxicology* 27. doi:10.5620/eht.2012.27.e2012018.

Sklar, Richard L. 1987. "Developmental democracy." *Comparative Studies in Society and History* 29 (4): 686–714. doi:10.1017/S0010417500014845.

Smith, Kirk R., Alistair Woodward, Diarmid Campbell-Lendrum, Dave D. Chadee, Yasushi Honda, Qiyong Liu, Jane M. Olwoch, Boris Revich, and Rainer Sauerborn. 2014. "Human health: impacts, adaptation, and co-benefits." In *Climate Change 2014: Impacts, Adaptation, and Vulnerability*, edited by C. B. Field, V. R. Barros, D. J. Dokken, K. J. Mach, M. D. Mastrandrea, T. E. Bilir, M. Chatterjee, K. L. Ebi, Y. O. Estrada, R. C. Genova, B. Girma, E. S. Kissel, A. N. Levy, S. MacCracken, P. R. Mastrandrea and L. L. White, 709–754. Cambridge/New York: Cambridge University Press.

Szreter, Simon, and Michael Woolcock. 2004. "Health by association? Social capital, social theory, and the political economy of public health." *International Journal of Epidemiology* 33 (4): 650–667. doi:10.1093/ije/dyh013.

Tompkins, Emma L., and Hallie Eakin. 2012. "Managing private and public adaptation to climate change." *Global Environmental Change* 22 (1): 3. doi:10.1016/j. gloenvcha.2011.09.010.

Tompkins, Emma L., W. Neil Adger, Emily Boyd, Sophie Nicholson-Cole, Keith Weatherhead, and Nigel Arnell. 2010. "Observed adaptation to climate change: UK evidence of transition to a well-adapting society." *Global Environmental Change* 20 (4): 627–635. doi:10.1016/j.gloenvcha.2010.05.001.

Tosun, Jale. 2013. *Risk Regulation in Europe – Assessing the Application of the Precautionary Principle*. New York: Springer.

TRAC3. n.d. *"About TRAC3."* Accessed 3 February 2019. http://trac3.ca/about-trac3/.

Tversky, Amos, and Daniel Kahneman. 1992. "Advances in prospect theory: cumulative representation of uncertainty." *Journal of Risk and Uncertainty* 5 (4): 297–323.

UNESCAP. n.d. *What is Good Governance?*United Nations Economic and Social Commission for Asia and the Pacific.

UNFCCC. n.d. "Parties & observers." Accessed 30 June 2019. https://unfccc.int/pa rties-observers.

Wæver, Ole. 2011. "Politics, security, theory." *Security Dialogue* 42 (4–5): 465–480. doi:10.1177/0967010611418718.

Watts, Nick, Markus Amann, Nigel Arnell, Sonja Ayeb-Karlsson, Kristine Belesova, Helen Berry, Timothy Bouley, Maxwell Boykoff, Peter Byass, Wenjia Cai, Diarmid Campbell-Lendrum, Jonathan Chambers, Meaghan Daly, Niheer Dasandi, Michael Davies, Anneliese Depoux, Paula Dominguez-Salas, Paul Drummond, Kristie L. Ebi, Paul Ekins, Lucia Fernandez Montoya, Helen Fischer, Lucien Georgeson, Delia Grace, Hilary Graham, Ian Slava Hamilton, Stella Hartinger, Jeremy Hess, Ilan Kelman, Gregor Kiesewetter, Tord Kjellstrom, Dominic Kniveton, Bruno Lemke, Lu Liang, Melissa Lott, Rachel Lowe, Maquins Odhiambo Sewe, Jaime Martinez-Urtaza, Mark Maslin, Lucy McAllister, Jankin Mikhaylov, James Milner, Maziar Moradi-Lakeh, Karyn Morrissey, Kris Murray, Maria Nilsson, Tara Neville, Tadj Oreszczyn, Fereidoon Owfi, Olivia Pearman, David Pencheon, Steve Pye, Mahnaz Rabbaniha, Elizabeth Robinson, Joacim Rocklöv, Olivia Saxer, Stefanie Schütte, Jan C. Semenza, Joy Shumake-Guillemot, Rebecca Steinbach, Meisam Tabatabaei, Julia Tomei, Joaquin Trinanes, Nicola Wheeler, Paul Wilkinson, Peng Gong, Hugh Montgomery, and Anthony Costello. 2018. "The 2018 report of the *Lancet* countdown on health and climate change: shaping the health of nations for centuries to come." *The Lancet* 392 (10163): 2479–2514. doi:10.1016/S0140-6736(18)32594-7.

Wei, Junni, Alana Hansen, Ying Zhang, Hong Li, Qiyong Liu, Yehuan Sun, and Peng Bi. 2014. "Perception, attitude and behavior in relation to climate change: a survey among CDC health professionals in Shanxi province, China." *Environmental Research* 134: 301–308. doi:10.1016/j.envres.2014.08.006.

Wendt, Alexander. 1992. "Anarchy is what states make of it: the social construction of power politics." *International Organization* 46 (2): 391–425.

White-Newsome, Jalonne L., Sabrina McCormick, Natalie Sampson, Miatta A. Buxton, Marie S. O'Neill, Carina J. Gronlund, Linda Catalano, Kathryn C. Conlon, and Edith A. Parker. 2014. "Strategies to reduce the harmful effects of extreme heat events: a four-city study." *International Journal of Environmental Research and Public Health* 11 (2): 1960–1988. doi:10.3390/ijerph110201960.

Whiteside, Kerry H. 2006. *Precautionary Politics – Principle and Practice in Confronting Environmental Risk*. Cambridge/London: MIT Press.

Wigley, Simon, and Arzu Akkoyunlu-Wigley. 2011. "The impact of regime type on health: does redistribution explain everything?" *World Politics* 63 (4): 647–677. doi:10.1017/S0043887111000177.

Wilkinson, P., D. H. Campbell-Lendrum, and C. L. Bartlett. 2003. "Monitoring the health effects of climate change." In *Climate Change and Human Health – Risks and Responses*, edited by A. J. McMichael, D. H. Campbell-Lendrum, C. F. Corvalán, K. L. Ebi, A. K. Githeko, J. D. Scheraga and A. Woodward, 204–219. Geneva: World Health Organization.

Wilkinson, Richard, and Michael Marmot. 2003. *Social Determinants of Health: The Solid Facts.* Vol. 2. Copenhagen: WHO.

Williams, Michael C. 2003. "Words, images, enemies: securitization and international politics." *International Studies Quarterly* 47 (4): 511–531.

Wise, Marilyn, and Peter Sainsbury. 2007. "Democracy: the forgotten determinant of mental health." *Health Promotion Journal of Australia* 18 (3): 177–183.

Wise, R. M., I. Fazey, M. Stafford Smith, S. E. Park, H. C. Eakin, E. R. M. Archer Van Garderen, and B. Campbell. 2014. "Reconceptualising adaptation to climate change as part of pathways of change and response." *Global Environmental Change* 28: 325–336. doi:10.1016/j.gloenvcha.2013.12.002.

Wurster, Stefan. 2011. "Sustainability and regime type. Do democracies perform better in promoting sustainable development than autocracies?" *Zeitschrift für Staats- und Europawissenschaften* 9 (4): 538–559.

Part II
The global perspective

4   Research design

The vast majority of research projects on adaptation to climate change utilizes either quantitative or qualitative methods and approaches to answer such research questions. Each method has its own advantages and limitations. Quantitative approaches and methods have proven very useful when looking for generalizations through comparing numerous cases, they show deficits with regards to understanding the causal relations behind different observations (Creswell and Clark 2011, 8). That is exactly where qualitative approaches and methods can demonstrate their strengths: due to the low number of units of investigation, they are more suitable for in-depth analysis of the causes and effects of certain observations (Collier, Brady, and Seawright 2010, 158). The detail-orientation does, however, come with a price: due to the small number of cases, it is usually difficult to generalize the results. Consequently, if combined properly, quantitative and qualitative methods can balance each other's weaknesses whilst making use of their individual strengths.

This study seeks to discover whether and how states adapt to climate change related health risks and factors that may explain the variance between their performances. To achieve this, a combination of quantitative and qualitative methods is necessary, since the quantitative part helps to identify general trends and the qualitative part contributes to a more detailed and in-depth understanding of the causal relations. While this conclusion might sound plausible to many readers, a great number of scholars have debated over years on whether quantitative and qualitative traditions can be combined or not. Therefore, the following sections will present the development of the academic discourse on the differences and similarities of quantitative and qualitative traditions and explain in greater detail how and why qualitative and quantitative methods will be combined in this study.

Quantitative vs. qualitative traditions

Although social science research, especially in Europe, frequently used a combination of quantitative and qualitative methods at the beginning of the twentieth century, mixing methods has subsequently been the subject of intense debates (Kuckartz 2014, 27).[1] Numerous scholars have argued that the respective perspectives and methods do not only differ with regards to their units of investigation but also concerning their underlying ontological, epistemological, and methodological foundations (Wolf 2015, 487).[2] Often, quantitative methods were perceived as directly linked to positivist[3] or postpositivist worldviews, as Creswell and Clark (2011, 40) describe it, whereas qualitative methods were associated with constructivist world views. Both worldviews (or paradigms)[4] follow different logics of research: postpositivism is, according to Creswell and Clark (2011, 40), based on "determinism or cause-and-effect thinking," "reductionism, by narrowing and focusing on select variables to interrelate," "detailed observations and measures of variables," and "the testing of theories that are continually refined." In contrast to postpositivism, constructivists follow the conviction that instead of determining the causes and effects of certain events and processes, scholars should do their best to understand the causal relations of various factors which may influence and be influenced by the dependent variable through social interaction (Creswell and Clark 2011, 40). Therefore, rather than determining top down and based on theoretical assumptions of which factors lead to which outcomes, constructivists believe that research is shaped "from the bottom up" and develops hypotheses and theories through the detailed analysis of empiricism (Creswell and Clark 2011, 40). The distinction of deduction (theory testing with empirical findings) versus induction (theory building based on empirical findings) is often wedded with the differences between postpositivists and constructivists and therefore also between quantitative and qualitative methods (Morgan 2007, 70).

It is, however, important to note that the understanding of the differences and similarities between postpositivist and constructivist worldviews and quantitative and qualitative methods may differ from scholar to scholar, depending on differences between their ontological and epistemological stances. Moreover, postpositivism and constructivism are by far not the only existing worldviews in social science research, but they rather constitute prominent and frequently discussed examples of debates on how to conduct research (Phoenix et al. 2013).

Furlong and Marsh (2010, 210) postulated that the ontological and epistemological beliefs behind the methodological differences of quantitative and qualitative methods were "skins, not sweaters" and could therefore not be used by the same researcher in different projects or even in the same research project. Accordingly, a great number of authors, including Mastenbroek and Doorenspleet (2007, 10) claimed that mixing methods was only possible under the same paradigm (Wolf 2015, 487). The term paradigm was first introduced in social science research by Kuhn (1970) in his book *The Structure of Scientific Revolutions*. Morgan (2007, 50), however, claims that Kuhn was quite unclear about the definition of a paradigm and actually used more than 20 different interpretations of the term throughout his book. Therefore, Morgan (2007, 51) distinguishes paradigms into four different versions: paradigms as worldviews, as epistemological stances, as shared beliefs in a research field, and as model examples. He claims that: "With regard to combining qualitative and quantitative methods, paradigms as epistemological stances have had a major influence on discussions about whether this merger is possible, let alone desirable" (Morgan 2007, 52). Moreover, he argues that how research questions are asked and answered depends on epistemological stances or so-called belief systems which social scientists have developed over time (Morgan 2007, 52). This controversial debate on the compatibility and possible superiority of specific methods is nowadays known as the so-called paradigm-wars (Lincoln and Guba 1985).

In an attempt to end the paradigm-wars, King, Keohane, and Verba (1994) wrote the seminal book *Designing Social Inquiry – Scientific Inference in Qualitative Research*. They intended to shatter the belief that the two research traditions or paradigms were fundamentally different. Their major point was that "the logic of good quantitative and good qualitative research designs do not fundamentally differ" (King et al. 1994, ix). As a consequence, they argued that both traditions could and should be combined.

Following their publication, King et al. sparked new debates in the research community, especially on what good research should look like (Collier et al. 2010, Tarrow 2010). Whilst a number of researchers appreciated King et al.'s efforts to establish stricter and more concise rules for good research, many accused them of trying to simply make qualitative research more like quantitative research (Collier et al. 2010, 125, Tarrow 2010, 101). The basis of this criticism is King et al.'s claim that the logic that underlies quantitative research should also be the foundation for the best qualitative research (Collier et al. 2010, 125, Tarrow 2010, 101). King et al. explicitly stated:

> Our main goal is to connect the traditions of what are conventionally devoted "quantitative" and "qualitative" research by applying a unified logic of inference to both. The two traditions appear quite different; indeed they sometimes seem to be at war. Our view is that these differences are mainly ones of style and specific technique. The same underlying logic provides the framework for each research approach.
>
> (King et al. 1994, 3)

Following King et al.'s publication, it took more than 15 years until Brady and Collier (2010) published the next groundbreaking book on quantitative and qualitative research, in which they synthesized debates on differences, similarities, and the potential compatibility of the two traditions. In *Rethinking Social Inquiry – Diverse Tools, Shared Standards*, they gave various scholars the chance to contribute to the debate on the differences and similarities of quantitative and qualitative traditions. A large share of the articles in Brady and Collier (2010) book reviews, praises, or criticizes King et al.'s statements and makes recommendations for how and why the paradigm-wars should be overcome (e.g. Collier et al. 2010, Tarrow 2010, King et al. 1994). Moreover, many of the articles focus on specific methods which have been designed to combine quantitative and qualitative perspectives (Collier et al. 2010, King, Keohane, and Verba 2010, Tarrow 2010).

Specific techniques to combine and mix methods have gained considerable attention in academia in recent years and have been applied by an increasing number of researchers (Collier et al. 2010, 127, Morgan 2007, 73). Most of them belong to one of the following categories: method combination, method integration, triangulation, and mixed methods designs (Kuckartz 2014, 29). Whilst the triangulation approach is rather popular in Germany and other European countries, in English-speaking countries a large number of scholars focuses on mixed methods (Kuckartz 2014, 29). To decide which specific research design is most suitable for the research goals of this study, the following section discusses the two approaches and their underlying epistemological and methodological foundations.

Combining methods – mixed methods research (MMR)

Mixed methods, multi methods, method triangulation – the number of terms and definitions for research that deals with the combination of different methods has continued to grow in recent years.[5] Johnson, Onwuegbuzie, and Turner (2007, 112) asked many prominent mixed methods researchers how they would define mixed methods research and their responses vary greatly. Teddlie and Tashakkori (2010, 8), however, "believe that there is a general agreement on some characteristics of […] MMR." The first general characteristic is called "methodological eclecticism" and means that MMR chooses the most appropriate methods for the research project from the great number of available qualitative (QUAL) and quantitative (QUAN) methods and integrates them into the research design (Teddlie and Tashakkori 2010, 9). Methodological eclecticism is more than "cancel[ling] out respective weaknesses of one or the other," but rather requires a researcher who is a "connoisseur of methods, who knowledgeably (and often intuitively) selects the best techniques available to answer research questions that frequently evolve as a study unfolds" (Teddlie and Tashakkori 2010, 9).

Teddlie and Tashakkori (2010, 9) argue that methodological eclecticism is based on the rejection of the "incompatibility of methods thesis," which was, among others, proclaimed by Furlong and Marsh (2010). According to Teddlie and Tashakkori (2010, 9), most MMR researchers follow the compatibility thesis and believe in the value of combining QUAN and QUAL methods. Therefore, the second characteristic of MMR they identify is "paradigm pluralism," which entails that a variety of different research traditions or paradigms can serve as the philosophy of MMR (Teddlie and Tashakkori 2010, 9).

The majority of researchers working with mixed methods builds their research on the concept of pragmatism, which was first introduced by Charles

Peirce's article "How to Make Our Ideas Clear" in 1878 (Kuckartz 2014, 42, Peirce 1878). The concept was further developed by philosophers, such as John Dewey and William James, and contemporary researchers, as for instance Cherryholmes (1992) or Murphy (1990) (see also Creswell and Clark 2011, 43). In contrast to other epistemological stances, pragmatism does not strictly abide by general beliefs about which ways of doing research are the best ones, but follows the assumption: "what works is what is useful and should be used, regardless of any philosophical assumption, or any other type of assumption" (Johnson and Christensen 2014, 491, Kuckartz 2014, 36).

The third characteristic of MMR, in the interpretation of Teddlie and Tashakkori (2010, 9), is "diversity at all levels," which entails that MMR can at the same time address various questions and both test existing hypotheses and develop new ones. This statement shows how enshrined the concept of pragmatism is in most MMR research. Pragmatists share different ontological, epistemological, and methodological worldviews from those of post-positivists and constructivists (Creswell and Clark 2011, 42). Their ontological under-standing is based on the assumption that both singular and multiple realities are possible, which results in the epistemological concept of practicality (Creswell and Clark 2011, 42). Practicality is characterized by the fact that researchers should conduct research that is centered around the research question and choose whether they remain distant and impartial or step closer into the research subject, depending on what works best to pursue the research goal (Creswell and Clark 2011, 42).

This goes along with the fourth characteristic, which states that, instead of dichotomies, the research is based on continua (Teddlie and Tashakkori 2010, 10). This allows the research project to not only look for either-or explanations, but explore a range of options from the whole epistemological and methodological spectrum (Teddlie and Tashakkori 2010, 10). Similarly, MMR allows the combination of both inductive and deductive logics in the same research project (fifth characteristic of MMR) and puts the research question or problem at the forefront of the research project (sixth char-acteristic of MMR) (Creswell and Clark 2011, 43, Teddlie and Tashakkori 2010, 10). Following the pragmatist stance, the research question should be regarded as the center of the research project and the methods used should be chosen according to the possibilities and challenges set by the research question (Creswell and Clark 2011, 43).

The seventh characteristic of contemporary MMR is centered around the research designs and analytical processes that MMR makes use of (Teddlie and Tashakkori 2010, 11).[6] Although a variety of names and interpretations of concrete research designs exists, most of them share some common traits. The key question is how the respective QUAN and QUAL methods are integrated into one coherent design (Johnson et al. 2007, 119–121). Follow-ing Wolf (2015, 490), the successful connection between the individual methods is what makes "multi methods research" to "mixed methods research." To achieve such a successful combination, useful connection points shall be identified so that the respective empirical parts (quantitative and qualitative part) can be connected (Morgan 2007, 71). There is not one perfect way to combine the respective parts, but what is important is a comprehensive discussion of the rationale behind the combination and the respective connection points for the mixed methods study (Wolf 2015, 491). Simply adding a few case studies to a quantitative study is not proper mixed methods research and does not make use of the potential benefits that the described approach offers (Wolf 2015, 491).

Among the variety of different traditions to combine methods, the most prominent approaches are called mixed methods and method triangulation (Flick 2011, 82–83, King et al. 2010, 122). Bearing in mind the high level of complexity of many research questions and topics, proponents of the mixed methods approach believe in the compatibility of quantitative and qualitative traditions and argue that both methods provide different perspectives from which research can benefit (Kuckartz 2014, 35). Responding to the long-lasting debates on the differences between quantitative and qualitative methods, proponents of mixed methods research and method triangulation have developed distinct ways and techniques of combining the underlying worldviews (Creswell and Clark 2011, 40–43).

At first glance, method triangulation and mixed methods look quite similar, especially regarding their pragmatic approach towards research, but on closer inspection some important differences become visible (Kuckartz 2014, 48). Triangulation simply means that a subject of interest is observed from two different perspectives (Wolf 2015, 483). Together with the subject of interest, the two perspectives construct a triangle which gives the approach its specific name (Wolf 2015, 483). Triangulation can mean the use of different data, researchers, theories, or methods to answer the same research question (Wolf 2015, 483). The reasons for using triangulation often rest in the belief that the approach results in more comprehensive, in-depth, and reliable findings than traditional approaches, which usually focus on only one perspective (Wolf 2015, 483). The same applies to mixed methods research and its distinct sub-forms, such as the explanatory sequential design (Creswell 2015, 7). Both triangulation and mixed methods seek to combine different methods and research approaches by overcoming potential obstacles concerning the underlying worldviews and focusing on how to best answer the research question (Creswell and Clark 2011, 43).

Whilst mixed methods and triangulation share many similarities, there are some major differences. First of all, the triangulation approach seeks to be independent from existing epistemological stances, whereas, according to Kuckartz (2014, 48), at least some parts of the mixed methods community perceive themselves as representatives of a new paradigm. Additionally, the triangulation approach constitutes a rather general and loose setting under which multiple methods can be combined, whereas in mixed methods research very specific options for combining methods exist (Kuckartz 2014, 48). In the beginnings of the triangulation approach, however, the most important difference was that the method was mainly used to validate information (Kuckartz 2014, 48).[7] Later, the founder of triangulation, Norman Denzin, denied such claims and stressed that triangulation should also be used to develop a deeper understanding of the research subject and not just to test hypotheses (Kuckartz 2014, 45).

Mixed methods, on the other hand, is frequently regarded as a more progressive concept that does not primarily focus on validation but on simply finding the perfect methods for answering the research question while adhering to fitting epistemological stances such as pragmatism (Kuckartz 2014, 50). Kuckartz (2014, 50) breaks the difference between mixed methods and triangulation down to the difference between a very general concept of validation and the enrichment of perspectives (triangulation) versus very specific forms of implementation and design (mixed methods). This claim is strengthened by the fact that three distinct and very specific designs of mixed methods research exist: the *convergent design*, the *explanatory sequential design*, and the *explanatory design* (Creswell 2015, 7). The convergent design collects both quantitative and qualitative data, analyses both data sets, and in the end merges the results in order to compare the results and enhance the quality of the study (Creswell 2015, 7). The explanatory sequential design begins with a quantitative analysis, which is then followed by a qualitative analysis to delve more into the potential causal relations behind the research subject (Creswell 2015, 7). Last, but not least, the explanatory design seeks to first explore a research subject with qualitative methods to further develop the research question and then use the qualitative results to draw broader and representative conclusions from a subsequent quantitative analysis (Creswell 2015, 7).

As a special form of the explanatory sequential design, in 2005, Lieberman introduced the Nested Analysis Approach as a "a unified approach which joins intensive case-study analysis with statistical analysis" (Lieberman 2005, 435). It usually starts with a preliminary Large-N-Analysis (LNA) to create an overview of global trends and correlations and to guide case selection for a subsequent Small-N-Analysis (SNA) that tests the results of the quantitative part, but also enables the qualitative part to build new models that can subsequently be tested through an additional LNA (Lieberman 2005, 435). In contrast to other mixed methods approaches, Lieberman's nested analysis follows the central question of whether the results of the LNA are robust and satisfactory (Lieberman 2005, 435). If yes, the study should proceed with a SNA of deliberate or random on-the-line cases to test whether the assumed independent variables can really explain the dependent variable (Lieberman 2005, 435). If the results from the LNA are not robust and satisfactory, the study should move on to a model-building SNA of deliberate "on-" and "off-the-line" cases (Lieberman 2005, 435). If the SNA leads to a new and coherent model, the study should test the model with a new LNA (Lieberman 2005, 435). If not, or if an additional LNA is not possible, the analysis should end (Lieberman 2005, 435).

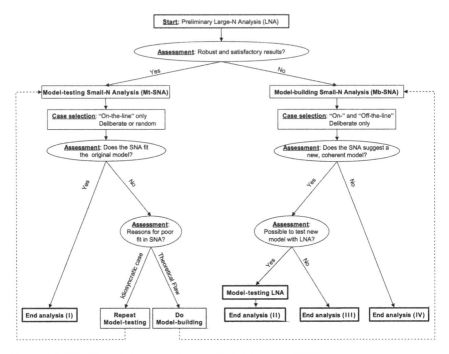

Figure 4.1 The Nested Analysis Approach: Lieberman (2005)

The nested analysis seeks to improve the methodological toolbox of making causal inferences in comparative research by combining quantitative and qualitative methods in an integrated approach (Lieberman 2005, 435). Despite the advantages of the nested analysis compared with other mixed methods designs, however, a number of authors have criticized that its approach favors quantitative over qualitative methods and accordingly does not dedicate as much time and thought into the qualitative parts of the analysis (Mastenbroek and Doorenspleet 2007, 16–17, Wolf 2015, 493). Although Lieberman speaks neither in favor nor against one or the other method, many studies that use the Nested Analysis Approach actually focus more on the LNA than the SNA (Lieberman 2005, 436, Tosun 2013, Wolf 2015, 493). Moreover, Rohlfing (2007, 1497) claims that all elements of causal inference of Lieberman's Nested Analysis Approach are "affected by ontological misspecification." Therefore, he claims that a study should begin all over again with completely new case studies if the results of the LNA are not satisfactory, instead of on- and off-the-line case selection (Rohlfing 2007, 1497).[8] Following Lieberman's suggestion, this study starts with a first LNA to investigate general trends and will continue with a theory-building SNA and a final, concept-testing LNA.[9] For the sake of readability, the two quantitative parts were combined in this book, but the original regression results can be found in the supplementary material.

Notes

1 According to Kuckartz (2014, 27), prominent authors, such as Max Weber, have frequently relied on a combination of qualitative and quantitative methods in their research projects.

2 Whilst many different interpretations of ontology, epistemology, and methodology exist in political science, this thesis follows the following definitions, which are based on Bates and Jenkins (2007), Blaikie (2012), Creswell and Clark (2011), and Hay (2011): 1. ontology describes the overall understanding of reality and especially a shared understanding of what exists and how, 2. epistemology, in contrast, is about the understanding of how to best conduct research and the underlying assumptions that accompany this understanding, such as what is knowledge and whether or how it can be produced; and 3. methodology is about the process of doing research, such as the analytical strategy and concrete tools which serve to achieve the research goal (Creswell and Clark 2011, 41).

3 While positivism follows a "hard science" approach in which a single reality exists and can be identified, postpositivism, which has in many areas become the successor of positivism, acknowledges that humans can never fully understand and identify reality, but can only approximate to what nature really is (Phoenix et al. 2013, 221).

4 For an explanation of the term paradigm see the following page.

5 For an overview of different MMR traditions and purposes, see Venkatesh, Brown, and Bala (2013).

6 Teddlie and Tashakkori (2010) added an eighth characteristic of contemporary MMR, "a tendency toward balance and compromise," and a ninth characteristic, "a reliance on visual representations," which will not be discussed in greater detail at this point since their value for this study is rather limited.

7 According to Kuckartz (2014, 48) this is at least the case for the traditional triangulation approach developed by (Denzin 1978).

8 Since Lieberman's model does not include the option of doing case-study based nested analyses and rather focuses on the quantitative part of the research design, Rohlfing (2007) and (Rohlfing and Starke 2013) introduce the concept of case-study based nested analysis.

9 Grounded theory, in the original understanding of Glaser and Strauss (2017) implies that researchers approach a research topic completely free of preset ideas on how to categorize or theorize the topic, but rather directly delve into the research subject and identify patterns in texts or other data sets, which ultimately lead to hypotheses and theoretical concepts (Strauss and Corbin 1994, 1997).

References

Bates, Stephen R., and Laura Jenkins. 2007. "Teaching and learning ontology and epistemology in political science." *Politics* 27 (1): 55–63.

Blaikie, Norman. 2012. *Approaches to Social Enquiry: Advancing Knowledge*. Cambridge; Malden, MA: Polity Press.

Brady, Henry, and David Collier. 2010. *Rethinking Social Inquiry – Diverse Tools, Shared Standards*. Plymouth: Rowman & Littlefield.

Cherryholmes, Cleo H. 1992. "Notes on pragmatism and scientific realism." *Educational Researcher* 21: 13–17.

Collier, David, Henry Brady, and Jason Seawright. 2010. "Critiques, responses, and trade-offs: drawing together the debate." In *Rethinking Social Inquiry – Diverse Tools, Shared Standards*, edited by Henry Brady and David Collier, 125–159. Plymouth. Rowman & Littlefield.

Creswell, John. 2015. *A Concise Introduction to Mixed Methods Research*. London and New York: Sage.

Creswell, John, and Vicki L. Plano Clark. 2011. *Designing and Conducting Mixed Methods Research*. 2nd ed. London and New York: Sage.

Denzin, Norman K. 1978. "Triangulation: a case for methodological evaluation and combination." *Sociological Methods*: 339–357.

Flick, Uwe. 2011. *Triangulation - Eine Einführung*. 3rd ed. Wiesbaden: Verlag Springer.

Furlong, Paul, and David Marsh. 2010. "A skin not a sweater: ontology and epistemology in political science." In *Theory and Methods in Political Science*, edited by David Marsh and Gerry Stoker, 184–211. Basingstoke/New York: Palgrave Macmillan.

Glaser, Barney G., and Anselm L. Strauss. 2017. *Discovery of Grounded Theory: Strategies for Qualitative Research*. London and New York: Routledge.

Hay, Colin. 2011. "Political ontology." In *The Oxford Handbook of Political Science*, edited by E. Goodin Robert. Oxford: Oxford University Press.

Johnson, R. Burke, Anthony J. Onwuegbuzie, and Lisa A. Turner. 2007. "Toward a Definition of Mixed Methods Research." *Journal of Mixed Methods Research* 1 (2): 112–133.

Johnson, R., and Larry Christensen. 2014. *Educational Research Quantitative, Qualitative, and Mixed Approaches* 5th ed. London: Sage.

King, Gary, Robert O. Keohane, and Sidney Verba. 1994. *Designing Social Inquiry – Scientific Inference in Qualitative Research*. Princeton. Princeton University Press.

King, Gary, Robert O. Keohane, and Sidney Verba. 2010. "The importance of research design." In *Rethinking Social Inquiry – Diverse Tools, Shared Standards*, edited by Henry Brady and David Collier, 111–122. Plymouth: Rowman & Littlefield.

Kuckartz, Udo. 2014. *Mixed Methods – Methodologie, Forschungsdesigns und Analyseverfahren*. Wiesbaden: Verlag Springer.

Kuhn, Thomas S. 1970. The structure of scientific revolutions. 2nd ed., *International Encyclopedia of Unified Science*. Chicago: University of Chicago Press.

Lieberman, Evan S. 2005. "Nested analysis as a mixed-method strategy for comparative research." *American Political Science Review* 99 (3): 435–452. doi:10.1017/S0003055405051762.

Lincoln, Yvonna S., and Egon G. Guba. 1985. *Naturalistic Inquiry*. London: Sage.

Mastenbroek, Ellen, and Renske Doorenspleet. 2007. "*Mind the gap! On the possibilities and pitfalls of mixed methods research.*" 4th ECPR General Conference.

Morgan, David. 2007. "Paradigms lost and pragmatism regained – methodological implications of combining qualitative and quantitative methods." *Journal of Mixed Methods Research* 1 (1): 48–76.

Murphy, John P. 1990. *Pragmatism: From Peirce to Davidson*. Boulder, CO: Westview Press.

Peirce, Charles. 1878. "How to make our ideas clear." *Popular Science Monthly* 12: 286–302.

Phoenix, Cassandra, Nicholas J. Osborne, Clare Redshaw, Rebecca Moran, Will Stahl-Timmins, Michael H. Depledge, Lora E. Fleming, and Benedict W. Wheeler. 2013. "Paradigmatic approaches to studying environment and human health: (forgotten) implications for interdisciplinary research." *Environmental Science & Policy* 25: 218–228. doi:10.1016/j.envsci.2012.10.015.

Rohlfing, Ingo. 2007. "What you see and what you get: pitfalls and principles of nested analysis in comparative research." *Comparative Political Studies* 41 (11): 1492–1514. doi:10.1177/0010414007308019.

Rohlfing, Ingo, and Peter Starke. 2013. "Building on solid ground: robust case selection in multi-method research." *Swiss Political Science Review* 19 (4): 492–512. doi:10.1111/spsr.12052.

Strauss, Anselm, and Juliet Corbin. 1994. "Grounded theory methodology." *Handbook of Qualitative Research* 17: 273–285.

Strauss, Anselm, and Juliet M.Corbin. 1997. *Grounded Theory in Practice*: London: Sage.

Tarrow, Sidney. 2010. "Bridging the quantitative-qualitative divide." In *Rethinking Social Inquiry – Diverse Tools, Shared Standards*, edited by Henry Brady and David Collier, 101–110. Plymouth: Rowman & Littlefield.

Teddlie, Charles, and Abbas Tashakkori. 2010. "Overview of contemporary issues in mixed methods research." In *Sage Handbook of Mixed Methods in Social & Behavioral Research*, edited by Abbas Tashakkori and Charles Teddlie, 1–41. Thousand Oaks, CA: Sage.

Tosun, Jale. 2013. *Risk Regulation in Europe – Assessing the Application of the Precautionary Principle*. New York: Springer.

Venkatesh, Viswanath, Susan A. Brown, and Hillol Bala. 2013. "Bridging the qualitative–quantitative divide: guidelines for conducting mixed methods research in information systems." *MIS Quarterly*, 37 (1): 21–54.

Wolf, Frieder. 2015. "Methodentriangulation." In *Handbuch Policy Forschung*, edited by Georg Wenzelburger and Reimut Zohlnhöfer, 483–504. Wiesbaden: Verlag Springer.

5 The CHAIn index

Among the major contributions of this doctoral thesis is the creation of the Climate Change and Health Adaptation Index (**CHAIn**). It ranks the health adaptation initiatives of 192 UN Member States and distinguishes numerous categories that help to compare the performances of states. The index scores rest on a comprehensive qualitative content analysis of a variety of official international documents of each state, such as the most recent National Communications (NCs) to the UNFCCC, National Adaptation Plans, and National Health Adaptation Plans. It classifies the health adaptation initiatives based on the level of adaptation, the type of adaptation, and the health risks that are addressed. Following the mixed methods design of this study, the index results are used for a multiple regression analysis to test whether major geographical, socioeconomic, and political factors influence the performance of different states. Based on the statistical analysis, five cases will be selected and analyzed in greater detail to contribute to a better understanding of the drivers of and barriers to health adaptation to climate change.[1] The results of the qualitative assessment will then be tested on a quantitative basis again. But first, this chapter elaborates on the logic behind the individual components of the index and explains how it works.

General concept

The CHAIn is based on the influential article "National-level Factors Affecting Planned, Public Adaptation to Health Impacts of Climate Change" by Lesnikowski et al. (2013), which introduced the Adaptation Response Score (ARS) and the Health Areas Score (HAS) to compare the adaptation initiatives of states to climate change related health risks. The ARS measures the "range of types of action being taken within each country" and the HAS "the range of health vulnerabilities being addressed at the groundwork and adaptation levels" (Lesnikowski et al. 2013, 1157). Lesnikowski et al.'s (2013) categories therefore help to identify which health risks are addressed by the states and what kind of adaptation they pursue to counter those risks. Their categories to measure the adaptation initiatives of states have been further refined and amended by Austin et al. (2016). Since the literature on the health effects of climate change is rapidly developing and new insights on tracking adaptation have been gained since 2016, the existing frameworks have been further refined and adjusted for this study to meet the needs of this research project and reflect the current state of research. The following paragraphs explain the similarities and differences between the existing work on tracking health adaptation and the CHAIn index. They also expound how the index is composed and how the results need to be interpreted.

Categories and scores

Following Lesnikowski et al. (2013) and Austin et al. (2016), the analytical framework of this doctoral thesis distinguishes between Levels of Action, Types of Action, and different Health Risks.[2] Levels of Action include recognition, groundwork, and adaptation (see Table 5.1).[3] In general, the Levels of Action describe how active states are and which general measures they take to respond to climate change related health risks.

Table 5.1 Levels of action (based on Austin et al. 2016: 6, Lesnikowski et al. 2013: 1155, and own reflections)

Level of Action	Description	Examples
Recognition	Recognition of the impact of climate change on health and/or expression of the need to adapt to climate change related health risks.	General statements on the impact of climate change on the health of the country's population and/or general statements that action is necessary.
Groundwork	These initiatives seek to build adaptive capacity, prepare for adaptation actions, or enable adaptation actions.	Impact and vulnerability assessments, research and development of conceptual tools or communication tools, stakeholder networking, policy recommendations, and/or strategies and plans.
Adaptation	Adaptation initiatives indicate that concrete action has been taken to reduce the health vulnerability of specific populations or to increase resilience. They seek to alter institutions, policies, programs, and practices or to build environments and mandates in order to respond to current or future climate change related health risks.	Legislative changes, departmental development (working groups, ministries, departments), public awareness and outreach, surveillance and monitoring, infrastructure and technology, program or policy evaluations, financial support for adaptation, medical interventions, and/or change of working practices and habits.

As Table 5.1 shows, recognition means that states acknowledge that climate change is related to certain health risks or generally has an impact on health. For the analysis, this means that whenever a statement that addresses the general impact of climate change on health or the need to act against climate change related health risks appears, the passage will be encoded as recognition. Recognition measures do not specify how adaptation measures should look, but rather express that adaptation is necessary. Consequently, the recognition category helps to better understand states' perception of climate change related health risks and how important such risks are for them, but it does not indicate how they respond to such risks.

Groundwork initiatives contain all actions that states take to prepare adaptation actions, such as research and development, strategies and plans, composing lists of recommendations of what could and should be done to adapt to climate change related health risks, or developing communication tools. In contrast to adaptation initiatives, groundwork measures do not indicate whether physical, financial, or policy measures have been adopted or implemented, but rather constitute measures that seek to prepare such actions. Adaptation initiatives mark that states have already taken tangible actions to adapt to climate change related health risks through legislative changes, infrastructure measures, financial instruments, capacity building, behavioral changes, or other ways.

Groundwork and adaptation initiatives are further distinguished into different types of adaptation (the category "Types of Action"). As a consequence, it is possible to precisely distinguish between the initiatives that states have adopted and thus better compare how they differ in their responses to climate change related health risks. For this study, the analytical frameworks of Lesnikowski et al. (2013) and Austin et al. (2016) have been used and updated. The reasons for adjusting the existing frameworks were a) to tailor the categories to the research question and needs of this study; b) to develop even clearer and more comprehensible categories to better distinguish them from one another; and c) to reflect the current state of research on climate change and health. The existing analytical frameworks have led to great advancements and important insights on states' adaptation initiatives, however, for the Large-N-Analysis (LNA) used in this study, the categories needed to be adjusted to the database and updated to mirror the most recent academic findings on the nexus between climate change and health.

As Table 5.2 shows, groundwork initiatives include the following categories: Research and Development, Recommendations, Strategies and Plans, and Communication. The category "Research and Development" describes state-sponsored research projects or vulnerability assessments on climate change related health risks. Climate vulnerability and capacity assessments, which analyze the specific health risks and the capacity to counter those risks, are in many cases the precondition for successful adaptation actions since they help countries to better understand what they can and cannot do

Table 5.2 Typology of adaptation initiatives (based on Austin et al. 2016, Biagini et al. 2014: 104, Lesnikowski et al. 2013, McMichael 2013)

Adaptation type	Description	Example	Similar classification in literature
Groundwork			
Research and Development (r&d)	Research on the impact of climate change on health and on possibilities to adapt to the risks posed by climate change.	Research projects on the impact of climate change on health, vulnerability and adaptation assessments.	Warning or observing systems (Austin et al. 2016); Research and development (Smit et al. 2000, Carter et al. 1994); Impact assessment/vulnerability assessment (Lesnikowski et al. 2013).
Recommendations (rec)	Recommendation of policies, based on research and development.	Proposals for funding, recommendations for the implementation of early warning, monitoring, communication etc.; Review of existing systems and recommendations on improvements.	Recommendation (Lesnikowski et al. 2013).
Strategies and Plans (str)	Strategies, framework documents, and specific adaptation plans of governments, governmental institutions, and governmental agencies.	Adaptation plans, livelihood diversification, drought planning, coastal planning, ecosystem-based planning, changing natural resource management (all with focus on health).	Administrative/institutional/organizational (Smit and Skinner 2002, Wilbanks and Kates 1999, Smit et al. 2000, Carter et al. 1994, Tompkins et al. 2010); Conceptual tools (Lesnikowski et al. 2013); Management and planning (Biagini et al. 2014).
Communication (com)	Systems for communicating climate information to help build resilience towards climate impacts (other than communication for early warning systems).	Decision support tools, communication tools, data acquisition efforts, digital databases, remote communication technologies.	Infrastructural/structural (Smit et al. 2000, Carter et al. 1994); Educational/informational (Smit and Skinner 2002, Wilbanks and Kates 1999, Smit et al. 2000, Carter et al. 1994); Information (Biagini et al. 2014).

Adaptation type	Description	Example	Similar classification in literature
Adaptation			
Policy (pol)	The creation of new policies or revisions of policies or regulations to allow flexibility to adapt to changing climate.	Mainstreaming adaptation into development policies, land-use specific policies, improvement of water resource governance, revised design parameters, ensuring compliance with existing regulations.	Legislative/Legal (Smit et al. 2000, Carter et al. 1994); Legislation (Lesnikowski et al. 2013).
Capacity Building (cap)	Developing and enhancing human resources, institutions, and communities, equipping them with the capability to adapt to climate change.	Training/workshops for knowledge/skills development, identification of best practices, training materials, improvement of public health systems, and scaling up and training of health workers and ministerial departments for adaptation to climate change related health risks.	Educational/informational (Smit and Skinner 2002, Wilbanks and Kates 1999, Smit et al. 2000, Carter et al. 1994, Tompkins et al. 2010); Medical interventions (Lesnikowski et al. 2013).
Education and Public Awareness (edu)	Programs and actions to improve public awareness and knowledge of climate change related health risks, dissemination of information on how to prepare and prevent climate change related health risks on an individual level, education programs.	Information campaigns, workshops, curriculum development, consulting of citizens.	Public awareness/outreach (Lesnikowski et al. 2013).

Adaptation type	Description	Example	Similar classification in literature
Infrastructure (inf)	Any new or improved hard or soft physical infrastructure aimed at providing direct or indirect protection from climate hazards.	Climate-resilient buildings, reservoirs for water storage, irrigation systems, canal infrastructure, sea walls, revegetation, afforestation, woodland management, increased landscape cover.	Infrastructural/structural (Smit et al. 2000, Carter et al. 1994); Physical infrastructure/"Green" infrastructure (Biagini et al. 2014).
Practice and Behavior (pra)	Revisions or expansion of practices and on-the-ground behavior that are directly related to building resilience.	Soil/land management techniques, climate-resilient crops or livestock practices, post-harvest storage, rainwater collection, expanding integrated pest management and pest control, and immunization.	Behavioral (Smit and Skinner 2002, Wilbanks and Kates, 1999).
Monitoring and Warning (mon)	Implementation of new or enhanced tools and technologies for communicating weather and climate risks and for monitoring changes in the climate system.	Developing, testing, and deploying monitoring or warning systems (which are human health relevant; e.g. for extreme weather events which are used to prevent risks for public health or UV radiation), upgrading weather or hydrometrical services.	Warning or observing systems (Austin et al. 2016); Research and development (Smit et al. 2000, Carter et al. 1994); Surveillance and monitoring (Lesnikowski et al. 2013).
Financing (fin)	New financing or insurance strategies to prepare for future climate disturbances.	Insurance schemes, microfinance, contingency funds for disasters.	Financial (Smit and Skinner 2002, Wilbanks and Kates 1999, Smit et al. 2000, Carter et al. 1994); Market mechanisms (Smit et al. 2000, Carter et al. 1994).
Technology (tech)	Develop or expand climate-resilient technologies.	Technologies to improve water use or water access, solar energy capacity, biogas, water purification, solar salt production.	Technological (Smit and Skinner 2002, Wilbanks and Kates 1999, Smit et al. 2000, Carter et al. 1994).

against climate change related health risks (Hanna and McIver 2014, 190). Lesnikowski et al. (2013, 26 Supp. Material) created distinct sub-categories for vulnerability and impact assessments and other research activities in their assessment. Since vulnerability and impact assessments serve the same purpose as other research projects, this study combines them into the overall category "Research and Development," thereby seeking to create a cleaner analytical framework.

Following Austin et al. (2016), other categories that Lesnikowski et al. (2013) originally listed, such as modeling tools or conceptual tools, have been subsumed under broader categories, such as Research and Development, Strategies and Plans, or Communication tools.[4] Recommendations include all recommended actions that are part of the analyzed documents but have not been implemented yet, nor do they refer to when the recommended actions should be implemented. As a consequence, following Lesnikowski et al. (2013, 27 Supp. Material), they represent something that "should be done." Strategies and Plans, a newly created category, describes strategic documents and action plans on climate change related health risks, which contain clear timelines on when specific actions shall be implemented.[5] Neither Lesnikowski et al. (2013) nor Austin et al. (2016) list this comprehensive category, but rather use sub-forms such as "plans," or "guidebooks, frameworks and toolkits" (Austin et al. 2016, 5 Supp. Material). The category "Communication" mostly consists of communication and decision support tools that gather information for policymakers and other key actors and thus allow them to better develop adaptation actions. They do not include warning or observing systems.

Similar to groundwork initiatives, the adaptation type categories stem mostly from Lesnikowski et al. (2013) and Austin et al. (2016) and have been adjusted and updated based on additional seminal articles on climate change and health, such as McMichael (2013). Adaptation initiatives include Policy, Capacity Building, Education and Public Awareness, Infrastructure, Practice and Behavior, Monitoring and Warning, Financing, and Technology. "Policy" initiatives refer to policy and legal measures, such as laws, government regulations, directives and ministerial policies that the state has adopted to take action against climate change related health risks.

"Capacity Building" measures actively seek to improve and expand the adaptive capacity of states through departmental changes or the establishment of new departmental units on climate change and health, improvements of public health systems, scaling up and expanding health services in response to climate change, and much more. Moreover, training for public health workers and emergency workers through comprehensive programs, which teach them about the causes, effects, and possible responses to the health effects of climate change, can contribute to better risk preparedness (Hanna and McIver 2014, 190, McMichael 2013, 1341).

The category "Education and Public Awareness" contains educational programs, awareness campaigns, and other projects and programs that seek to raise attention for the different risks that climate change poses to health and explain what citizens can do to adapt to such risks. "Infrastructure" initiatives include infrastructure projects to reduce health risks associated with climate change, such as installing air conditioning systems or equipping buildings with pollen filters to reduce the impact of climate change on allergic reactions.

"Practice and Behavior" refers to measures that seek to alter existing practices related to public health, such as soil, land, or water management techniques.

"Monitoring and Warning" includes systems to warn the population against climate change related health risks, such as extreme weather events or infectious diseases. Investments in information, monitoring, and early-warning systems for extreme weather events may help both decision makers and the entire society to better grasp the challenges and respond to them (Hanna and McIver 2014, 190). Among many other measures, some countries have already successfully established official warning systems to inform the population in cases of exposure to higher levels UV radiation and the increased possibility of severe health risks (WHO/WMO 2012, 39).

"Financing" describes all financial instruments that aim at strengthening health adaptation measures, such as special funds or insurance schemes. Establishing emergency funds to restore acceptable conditions for affected populations after extreme weather events can, for instance, improve the adaptive capacity of states and the international community (Neira et al. 2008, 424). Such funds are of particular importance for small island development states which often lack the financial means of adapting to these risks themselves and therefore depend on international support (Hanna and McIver 2014, 190).

"Technology" constitutes the last category on the adaptation level and focuses on technological innovations that have been implemented to reduce climate change related health risks, such as special tools to improve access to clean water or smart urban design that helps to prevent heatwaves, such as apps to warn the population of extreme heat or humidity through a combination of sensors and weather forecasting.

The last typology that serves as a basis for the analytical framework of the CHAIn focuses on the specific health risks that are associated with climate change (the category "Health Risks"). Following Austin et al. (2016), McMichael (2013), as well as the literature review in the introduction of this book (Chapter 1), the categories are distinguished into general health risks as well as primary, secondary, and tertiary climate change related health risks. As Table 5.3 shows, primary health risks include heat- and cold-related diseases, health risks due to droughts and fires or floods and storms, as well as risks due to climate change related decreased air quality and negative effects of increased exposure to ultraviolet radiation, such as skin diseases or cataracts.

Table 5.3 Typology of climate change related health risks (based on Butler et al. 2014, Austin et al. 2016, McMichael 2013 and others)

Health risk		Description of risk
General health (gen)		General climate change related health risks without specific explanations of which risks are addressed.
Primary risks		
Extreme temperature	Heat-related (heat)	Heat cramps, heat edema, heat exhaustion, heat strokes, dehydration, and heat-related cardiovascular diseases or respiratory diseases.
	Cold-related (cold)	Death from exposure or cold-related cardiovascular diseases or respiratory diseases.
Extreme weather events	Droughts and fires (dro)	Cardiovascular diseases or respiratory diseases due to fire-related heat and fumes, burn injuries, dehydration, dying of thirst.
	Floods and storms (flo)	Drowning or freezing to death, injuries due to collapsing buildings or trees.
Air quality (air)		Decreased air quality as a consequence of exacerbated effects of air pollution due to climate change, more frequent respiratory diseases due to lacking access to fresh air because of climate change and more time spent indoors.
Ultraviolet radiation (uv)		Effects on the immune system, development of skin cancer and non-Hodgkin's lymphoma, cataracts.
Secondary risks		
Infectious disease (inf)		Diseases caused by pathogenic organisms, such as bacteria, viruses, parasites, or fungi. Can be transmitted directly or indirectly from one person to another. Zoonotic diseases are infectious diseases that can be spread from animals to humans.
Water and sanitation (wat)		Waterborne diseases as a consequence of destroyed infrastructure and lacking access to potable water and sanitation.
Food and agriculture (food)		Malnutrition, starvation, and foodborne diseases as a consequence of failed harvests due to droughts, fires floods, storms, extreme heat or cold, and other changing climate patterns.
Allergies (all)		Increased prevalence of allergic diseases due to allergenic pollen in the air and other factors.
Tertiary risks		
Mental health (men)		Mental health risks such as trauma, chronic stress, anxiety, or depressions can be triggered or enforced by the loss of properties or family members due to climate change related disasters; can be also caused by direct effects of increasing temperatures on mental health, especially on those already dealing with depressions, and many more.

Health risk	Description of risk
Migration (mig)	Climate change can have a major impact on various factors which can trigger and influence migration patterns, such as access to water and other resources (e.g. territory). Migration can have a severe impact on the health of migrants since access to health services, potable water, food, etc. is often restricted during the process of migration and sometimes even in host countries. As a consequence, migrants might suffer from a wide range of health issues, such as malnutrition, food- and waterborne diseases, dehydration, etc. Even mental health issues may arise because of what migrants experience on the way to their host countries.
Conflicts (con)	Climate change may lead to increasing violence and conflicts across the globe due to reduced availability and access to natural resources, water, habitable territories, etc. Violence again may drastically impact public health because people get hurt or even killed, may have limited access to health services and other pivotal goods for their health. Additionally, conflicts may spur migration, which again may have an influence on public health.

Secondary health risks include the effects of climate change on water and sanitation, food-related diseases and malnutrition, the spread of infectious diseases, as well as a higher prevalence of allergic reactions. Tertiary health risks are related to climate change, yet the connection is very diffuse, and it is often difficult to clearly state what the causing factors behind the health risks are. It is clear, however, that climate change has a negative impact on all these factors. They include mental health disorders as well as health risks as a consequence of climate change related migration or conflicts.

All primary, secondary, and tertiary health risks are only encoded as such if the states specifically mention them and link them to climate change. If risks or general adaptation measures are mentioned, but the nexus between climate change and health is not addressed, the initiatives are not encoded. This goes along with the definition of health adaptation to climate change in the introduction (Chapter 1). More details on how the initiatives are identified and encoded can be found in the codebook in the supplementary material.

CHAIn Country Score = recognition + groundwork + adaptation

Comparing health adaptation – the formula

Comparing the adaptation initiatives of different states is a complicated and complex endeavor, yet some groundbreaking academic articles have shown that a comparison is possible if researchers use transparent and reliable methods (Biesbroek et al. 2010, Lesnikowski et al. 2013, Austin et al. 2016). To ensure that the respective country scores are indeed comparable, and not arbitrary categories researchers put on the actual initiatives, the factors need to be carefully selected and weighted before creating the final index. Therefore, the CHAIn is composed of the above described categories, based on the respective levels of adaptation and the therewith included types of adaptation and addressed health risks.

The scores in the adaptation type categories are weighted to ensure an actual representation of the country's adaptation portfolio and to guarantee that all scores are comparable despite different styles and languages used in the official documents that compose the database of the assessments. Recognition scores are not counted, but as soon as the respective risks are mentioned in the document the country is awarded a point in the category. Recognition scores are weighted with the factor 0.5 to adjust for the relative impact of recognition initiatives compared with groundwork and adaptation initiatives. The rationale behind this is that countries that do not take action, but are aware that certain key words need to be included in official documents, do not receive an exponentially higher score than those countries that have actually initiated first adaptation initiatives but do not repeat the same risks multiple times throughout the documents.[6] Accordingly, recognition scores inform about the range of recognized risks states acknowledge, which helps to measure how comprehensive their understanding and perception of climate change related health risks is. It does, however, not show how seriously states take the risks or what they do against them.

recognition = 0.5 × (gen + heat + cold + dro + flo + air+ uv + inf + wat + food + all + men + mig + con)

Groundwork initiatives are counted and weighted with factors 0.25 (recommendation) and 0.5 (all other groundwork initiatives) to ensure that the overall scores are comparable. Without a difference in the factor, countries with a very high amount of recommendations would be much higher on the CHAIn index than countries that have already implemented concrete and tangible measures to adapt to climate change related health risks. This would thus lead to a distorted representation of the actual measures. Nevertheless, groundwork initiatives are of high importance for the overall adaptation portfolio since they help to prepare and develop tangible actions and thus need to be taken into account as well.

groundwork = 0.5 × (r&d + str + com) + 0.25 × rec

For the same reasons, all adaptation initiatives identified in the analyzed documents are counted and then weighted with the factor 2. After conducting numerous tests and comparisons of different weighting mechanisms it became clear that it is necessary to weight adaptation actions with the factor 2 and groundwork and recognition initiatives with factors 0.5 and lower since adaptation actions have a much higher impact on how well states are prepared for climate change related health risks. Concurrently, recognition and groundwork initiatives still matter and show how seriously states take climate change related health risks and how many risks they are aware of, even if they do not have the capacity to implement respective actions.

$$\textbf{adaptation} = 2 \times (\text{cap} + \text{edu} + \text{inf} + \text{pra} + \text{mon} + \text{fin} + \text{tech})$$

The described procedure for collecting and weighting the scores for each state has proven to be reliable, transparent, and well-functioning.[7] Most importantly, it allows us to compare the policies of a high number of states in a coherent and objective manner. Four different procedures of collecting and weighing the scores have been tested and the described method has proven to be the most effective and transparent one (see Appendix for other options). Compared with other weighting options, such as those using higher factors for groundwork initiatives or recognition initiatives, the rankings of states did not significantly change, which further demonstrates

Table 5.4 Scores, factors, and explanations of the CHAIn

Adaptation Type	Score	Factor	Explanation
Recognition	1 point for each health risk that has been addressed, independent of how often the risk has been mentioned.	0.5	The recognition score is weighted with the factor 0.5 to ensure that states with recognition scores in the majority of health categories, but a very low number of groundwork-level and adaptation-level initiatives, do not rank higher than those focusing on the major health risks by which they are affected by actually taking lots of actions. States receive a score in the recognition category as soon as they acknowledge that the respective health risk is related to climate change, yet it does not entail any action whatsoever. The CHAIn index seeks to compare how states respond to climate change related health risks, which is best displayed by their groundwork and adaptation initiatives. Therefore, states only get 1 point for each category and the mentions of the health risks are not counted. At the same time, these measures make sure that the recognition of specific health risks by the respective states can be observed and the recognition scores do not distort the overall index scores.

Adaptation Type	Score	Factor	Explanation
Groundwork			
Research and Development	Sum of all identified initiatives	0.5	Research and Development actions are weighted with the factor 0.5 to adjust for the impact of groundwork initiatives in relation to adaptation initiatives.
Recommendation	Sum of all identified initiatives	0.25	Recommendations are weighted with the factor 0.25 to adjust for the high number of recommendations that states make to prepare future strategies and adaptation actions. To ensure that all adaptation scores of states are comparable, despite different languages and styles used in their official documents, it is important to weight recommendations accordingly since some states with many recommendations but no adaptation actions would otherwise rank significantly higher than those that have responded to climate change related health risks with adaptation actions.
Strategies and Plans	Sum of all identified initiatives	0.5	Strategies and Plans are weighted with the factor 0.5 to adjust for the impact of groundwork initiatives in relation to adaptation initiatives.
Communication	Sum of all identified initiatives	0.5	Communication initiatives are weighted with the factor 0.5 to adjust for the impact of groundwork initiatives in relation to adaptation initiatives.
Adaptation			
Policy	Sum of all identified initiatives	2	All adaptation initiatives are weighted with the factor 2 to adjust for their relative impact compared with groundwork initiatives. Adaptation initiatives stand for already implemented actions to prevent climate change related health risks and therefore constitute stronger responses to such risks than groundwork initiatives.
Capacity Building	Sum of all identified initiatives	2	
Education and Public Awareness	Sum of all identified initiatives	2	
Infrastructure	Sum of all identified initiatives	2	
Practice and Behavior	Sum of all identified initiatives	2	
Monitoring and Warning	Sum of all identified initiatives	2	
Financing	Sum of all identified initiatives	2	
Technology	Sum of all identified initiatives	2	

the reliability and validity of the model. The chosen and here explained option takes into account the real-life impact of adaptation-level actions compared with recognition-level and groundwork-level measures and makes the country scores more comparable.

Data and codes

The database of the CHAIn consists of the most recent NCs to the UNFCCC, national adaptation or climate change strategies and plans, national climate change and health strategies, or official national climate change and health vulnerability assessments, if the respective states have adopted such documents.[8] In total, sufficient information for 192 countries was collected. For Equatorial Guinea, Western Sahara, and Libya, no NCs or adaptation documents were published until the end of the investigation period in May 2018, and these states therefore needed to be omitted from the database. To guarantee that the most recent publication was identified for each state, a structured online search for relevant climate change and adaptation documents has been conducted for every country.[9] The publication dates range from the year 2000 (first and most recent NC of Grenada) to 2018. While the large variance between the publication dates of the analyzed documents constitutes a challenge for comparing the adaptation policies of the countries, the fact that no more recent publications exist in some countries demonstrates that these take less action against climate change than those with more recent publications.[10] The documents were in Chinese, Czech, English, French, German, Greek, Italian, Lithuanian, Romanian, Russian, Slovak, Slovenian, Spanish, Swedish, and Ukrainian. The English, French, German, and Spanish documents constituted the database for 180 out of 192 countries and were analyzed by me. For the remaining 12 documents in other languages, native speakers were trained to conduct the assessment, and their results were again tested through an independent analysis by me, using Google Translate.

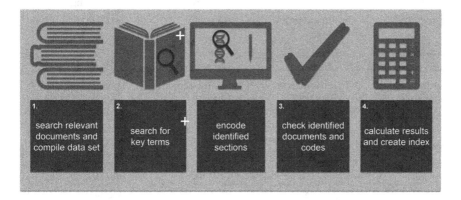

1.	2.		3.	4.
search relevant documents and compile data set	search for key terms	encode identified sections	check identified documents and codes	calculate results and create index

Figure 5.1 Coding instructions

As Figure 5.1 illustrates, the assessment started with a structured online search of the relevant documents and the compilation of the database. The data collection included: a) finding the most recent NC to the UNFCCC of the respective country; b) identifying whether national adaptation plans and strategies or general climate change strategies and plans exist and adding them to the database; c) identifying whether specific health adaptation strategies or plans to climate change or climate change risk assessments exist and adding them to the database. In addition to searching for the documents via Google and international organizations, such as the UNFCCC, UNDP, the EU, and others, all websites of Environmental Ministries and Health Ministries were checked for relevant documents.

The data collection was followed by a qualitative content analysis, using MAXQDA. After all relevant documents, together with the code system, had been imported into a MAXQDA file for the respective state, the search for key words began. Based on the categories from the CHAIn (level of adaptation, adaptation type, health risks) and precise coding instructions, the different terms and passages had been encoded to identify states' health adaptation initiatives. Table 5.5 summarizes the key words in English, Spanish, French, and German that were used in the analysis.

Following the analysis, all documents and codes were checked numerous times to avoid substantial overlaps, doubled codes, or missing codes. After

Table 5.5 Key words for the CHAIn analysis

English	Spanish	French	German
health	Salud	santé	Gesundheit
disease	Enfermedad	maladie	Krankheit
cardiovascular	cardiovascular	cardiovasculaire	Kardio-vaskular
respiratory	respiratoria	respiratoire	Respiratorisch/atem
dehydration	deshidratación	déshydratation	Dehydrieren, Verdursten
injury	lesión, daño	blessure	Verletzung
drowning	ahogamiento, ahogo	noyade	Ertrinken
freeze	frío	geler	Erfrieren
ultraviolet, skin	ultravioleta	ultraviolet	Ultraviolet, Haut
sanitation OR hygiene	saneamento, higiene	assainissement, hygiène	Hygiene
migration, displacement	migración, desplazamiento	migration, déplacement	Migration, Vertriebene
conflicts	conflictos	conflits	Konflikte
psychological, mental	psicológico	psychologique	Psychologische, mentale
allergies/allergy	alergia	allergies	Allergien

the results had been calculated for each state, they were transferred to the master spreadsheet of the CHAIn. Detailed information on the entire analysis, including all keywords and explanations, can be found in the codebook in the Appendix.

Index results

The CHAIn index (Table 5.6) shows that two states – the United Kingdom (UK) and the Republic of Korea (RoK) – score significantly higher than all other states, but both have different portfolios. While the UK has the highest possible recognition score (14 out of 14 climate change related health risks recognized) and benefits from very high groundwork scores, especially with regards to research and development, the RoK demonstrates by far the highest adaptation score of all states included in the index. At the same time, the RoK shows a medium recognition and a rather low groundwork score.

Following the RoK and the UK are Jordan, with the highest groundwork score of all countries, Canada, Cyprus, Moldova, the Solomon Islands, Germany, the Seychelles, and Sri Lanka. It clearly shows that the top ten states are very diverse with regards to their country and population size, economic performance, political systems, and many other factors. Moreover, numerous states that usually tend to perform well in prominent performance indices, such as the Gross Domestic Product (GDP) based on purchasing power parity (PPP), GDP (PPP) per Capita, or Transparency International's Corruption Perceptions Index (CPI), rank very low on the CHAIn. Prominent examples are Luxembourg (rank 165) or Iceland (last rank).

Table 5.6 Climate Change and Health Adaptation Index (CHAIn)

Rank	Country name	CHAIn score	Recognition	Ground-work	Adapta-tion
1	United Kingdom of Great Britain and Northern Ireland	109	7	42	60
2	Republic of Korea	103.25	5.5	9.75	88
3	Jordan	71.25	7	56.25	8
4	Canada	70.5	7	23.5	40
5	Cyprus	61	5.5	7.5	48
6	Republic of Moldova	50.25	6	20.25	24
7	Solomon Islands	49.25	4.5	30.75	14
8	Germany	47.5	6	27.5	14
9	Seychelles	46.25	4	22.25	20
10	Sri Lanka	46.25	5.5	12.75	28

Rank	Country name	CHAIn score	Recognition	Ground- work	Adapta- tion
11	Uganda	42.75	6.5	14.25	22
12	United States of America	42	4	4	34
13	Turkey	40	6	8	26
14	North Macedonia	37	5.5	31.5	0
15	Nepal	35.5	6.5	3	26
16	Netherlands	34	6.5	5.5	22
17	Belgium	33.75	5.5	6.25	22
18	Finland	33.75	6	7.75	20
19	Myanmar	32.5	5.5	9	18
20	Namibia	32.25	4.5	9.75	18
21	Russian Federation	31.25	5	16.25	10
22	China	30.25	3	11.25	16
23	United Republic of Tanzania	30	4.5	7.5	18
24	Singapore	29.25	4	5.25	20
25	Colombia	28.25	5	11.25	12
26	Italy	28.25	5.5	14.75	8
27	Saudi Arabia	27.5	5.5	4	18
28	France	27.25	6	11.25	10
29	Malawi	27.25	4	5.25	18
30	Botswana	27	1.5	1.5	24
31	Uruguay	26.25	5	7.25	14
32	New Zealand	25.75	6.5	11.25	8
33	Trinidad and Tobago	25.75	4.5	5.25	16
34	Rwanda	25.5	3.5	8	14
35	Australia	24.75	4	4.75	16
36	Denmark	24.75	5.5	7.25	12
37	Czech Republic	24.5	7	15.5	2
38	Switzerland	24.5	5	9.5	10
39	Mauritius	24.25	4	8.25	12
40	Brazil	24	5.5	6.5	12
41	Micronesia (Federated States of)	23.75	4.5	19.25	0
42	Bolivia	23.25	5.5	5.75	12
43	Jamaica	23.25	5.5	11.75	6
44	Philippines	22.75	4.5	6.25	12
45	Armenia	22.5	4.5	6	12
46	Sierra Leone	22.5	6	16.5	0
47	Spain	22.25	6	12.25	4

Rank	Country name	CHAIn score	Recognition	Ground-work	Adapta-tion
48	Timor-Leste	22	6	10	6
49	Ecuador	21.25	3.5	9.75	8
50	Malaysia	21	3	6	12
51	Chile	20.25	7	11.25	2
52	Egypt	20.25	5.5	8.75	6
53	Suriname	20.25	4.5	9.75	6
54	Norway	19.75	5	2.75	12
55	Fiji	19.5	5.5	10	4
56	Brunei Darussalam	19.25	2	1.25	16
57	Peru	19.25	5	6.25	8
58	Malta	18.5	5.5	11	2
59	Saint Vincent and the Grenadines	18.5	3.5	1	14
60	Lesotho	18	5	13	0
61	Uzbekistan	18	3	3	12
62	Venezuela (Bolivarian Republic of)	17.75	3	2.75	12
63	Comoros	17.5	5.5	12	0
64	Eritrea	17.5	3	10.5	4
65	Mexico	17.5	1.5	4	12
66	Belize	17	4.5	10.5	2
67	Guyana	17	5.5	5.5	6
68	Maldives	17	4	3	10
69	Sudan	17	4.5	12.5	0
70	Albania	16.75	6	6.75	4
71	Israel	16.5	4.5	10	2
72	Madagascar	15.75	2.5	9.25	4
73	Bangladesh	15.5	6.5	7	2
74	Greece	15.5	6	3.5	6
75	India	15.5	4.5	7	4
76	Austria	15.25	5.5	5.75	4
77	Ethiopia	15.25	5.5	9.75	0
78	Cambodia	15	4.5	10.5	0
69	Lebanon	15	5.5	9.5	0
80	South Africa	15	6	5	4
81	Lithuania	14	5.5	4.5	4
82	Togo	14	3.5	4.5	6
83	Democratic People's Republic of Korea	13.75	3	0.75	10
84	Iran (Islamic Republic of)	13.75	5.5	8.75	0

Rank	Country name	CHAIn score	Recognition	Ground-work	Adapta-tion
85	Japan	13.75	4.5	7.25	2
86	San Marino	13.5	4	3.5	6
87	Saint Lucia	13.25	3.5	5.75	4
88	Guatemala	13	3	6	4
89	Nauru	13	5	8	0
90	Serbia	13	5	0	8
91	Hungary	12.75	5	5.75	2
92	Ghana	12.5	4	4.5	4
93	Mali	12.5	2.5	2	8
94	Papua New Guinea	12.5	2	10.5	0
95	Portugal	12.5	5	1.5	6
96	Dominica	12.25	3.5	8.75	0
97	Samoa	12.25	5.5	6.75	0
98	Georgia	11.75	5.5	6.25	0
99	Kyrgyzstan	11.5	4.5	7	0
100	Turkmenistan	11.5	3.5	2	6
101	Côte D'Ivoire	11.25	4	7.25	0
102	Bosnia and Herzegovina	10.75	4.5	6.25	0
103	Liberia	10.75	4	2.75	4
104	Mongolia	10.75	5	5.75	0
105	Somalia	10.75	5	5.75	0
106	Bhutan	10.5	3	7.5	0
107	Burkina Faso	10.5	3.5	7	0
108	Estonia	10.5	6.5	4	0
109	Viet Nam	10.5	3	7.5	0
110	Honduras	10.25	4	6.25	0
111	Slovakia	10.25	5.5	4.75	0
112	Ukraine	10.25	4	6.25	0
113	El Salvador	10	4.5	5.5	0
114	Palestine	10	6	4	0
115	Costa Rica	9.75	3.5	2.25	4
116	Croatia	9.75	4	5.75	0
117	Central African Republic	9.5	2	7.5	0
118	Democratic Republic of the Congo	9.5	3.5	6	0
119	Poland	9.5	6	3.5	0
120	Kiribati	9.25	4	5.25	0
121	Niger	9.25	4.5	2.75	2
122	Sweden	9.25	4	5.25	0

Rank	Country name	CHAIn score	Recognition	Ground-work	Adapta-tion
123	Ireland	9	6.5	0.5	2
124	Paraguay	9	2	7	0
125	Zambia	8.5	3	5.5	0
126	Afghanistan	8.25	4.5	1.75	2
127	Bahamas	8.25	3	3.25	2
128	Gambia	8.25	5	3.25	0
129	Latvia	8.25	4.5	3.75	0
130	Slovenia	8.25	3	1.25	4
131	Burundi	8	4	4	0
132	Montenegro	8	2	4	2
133	Tajikistan	8	3.5	4.5	0
134	Cameroon	7.75	3.5	4.25	0
135	Iraq	7.75	5	2.75	0
136	Saint Kitts and Nevis	7.75	4.5	3.25	0
137	Bahrain	7.5	4	3.5	0
138	Romania	7.5	5	2.5	0
139	Angola	7.25	3.5	3.75	0
140	Benin	7.25	2.5	2.75	2
141	Argentina	7	4	3	0
142	Dominican Republic	7	3.5	3.5	0
143	Lao People's Democratic Republic	7	2.5	4.5	0
144	Liechtenstein	7	4.5	2.5	0
145	Nigeria	7	4	3	0
146	Swaziland	7	4	3	0
147	Indonesia	6.5	1.5	3	2
148	Tonga	6.5	2.5	4	0
149	Antigua and Barbuda	6.25	3.5	2.75	0
150	Kenya	6.25	3	3.25	0
151	Palau	6	3	1	2
152	Chad	5.75	3	2.75	0
153	Cuba	5.75	3.5	2.25	0
154	Syrian Arab Republic	5.75	3.5	2.25	0
155	Azerbaijan	5.5	1.5	4	0
156	Belarus	5.5	3.5	2	0
157	Grenada	5.5	4	1.5	0
158	Tuvalu	5.5	3.5	2	0
159	Marshall Islands	5.25	3	2.25	0
160	Nicaragua	5	1.5	1.5	2

Rank	Country name	CHAIn score	Recognition	Ground-work	Adapta-tion
161	Senegal	5	2	3	0
162	Tunisia	5	2	3	0
163	Vanuatu	5	4	1	0
164	Cabo Verde	4.75	3.5	1.25	0
165	Luxembourg	4.75	3	1.75	0
166	Morocco	4.75	2.5	0.25	2
167	Panama	4.75	3.5	1.25	0
168	Congo	4.5	3.5	1	0
169	Pakistan	4.5	2.5	0	2
170	Gabon	4	3.5	0.5	0
171	Guinea Bissau	4	2.5	1.5	0
172	Oman	4	0	0	4
173	Zimbabwe	4	2	2	0
174	Kuwait	3.75	1	0.75	2
175	Monaco	3.5	3	0.5	0
176	Haiti	3.25	1.5	1.75	0
177	Kazakhstan	3.25	1	0.25	2
178	Andorra	3	3	0	0
179	Mauritania	3	2	1	0
180	South Sudan	3	2.5	0.5	0
181	Thailand	3	0.5	2.5	0
182	Mozambique	2.75	2.5	0.25	0
183	United Arab Emirates	2.75	1.5	1.25	0
184	Djibouti	2.5	2.5	0	0
185	Yemen	2.5	2.5	0	0
186	Barbados	2.25	0	0.25	2
187	Guinea	2.25	1.5	0.75	0
188	Qatar	2.25	2	0.25	0
189	Sao Tome and Principe	1.75	1.5	0.25	0
190	Bulgaria	1.5	1	0.5	0
191	Algeria	0.5	0	0.5	0
192	Iceland	0	0	0	0

Recognition

The recognition index has led to a number of interesting, and sometimes partially surprising, findings. Although many states that perform well on the CHAIn index also had high recognition scores, a similar number of them do not follow this general trend. Some very high performing countries on the CHAIn, such as the Solomon Islands (rank 7), the Seychelles (rank 9), the United States (rank 12), Namibia (rank 20), or China (rank 22), have medium to low rankings on the recognition score, with China having the lowest of this group (recognition score: 3 out of 7). The recognition index shows the bandwidth of health risks recognized by the states and thus indicates which risks take priority for them. China, for instance, lists numerous general groundwork-level and adaptation-level initiatives and a few ones related to heat-related health risks, risks associated with extreme weather events, infectious diseases, and risks related to water and sanitation. Other climate change related health risks, such as those that are associated with higher exposure to UV radiation or related to food and agriculture, allergies, or tertiary health risks, are, however, not mentioned at all in the official national adaptation documents.

At the same time, numerous countries that rank medium to low on the CHAIn, such as Poland (rank 119), Ireland (rank 123), or Iraq (rank 135), show comparatively high recognition scores, yet they perform rather low when it comes to groundwork and adaptation initiatives (see Table 5.6). Ireland, for example, is placed on rank 123 on the CHAIn, with an overall score of 9. Compared with other states with a similar ranking, Ireland shows a very high recognition score (6.5), but at the same time low groundwork (0.5) and adaptation scores (2). This indicates that, despite Ireland's awareness of a great variety of climate change related health risks, the country is lagging behind in its adaptation actions.

Overall, 188 out of 192 states acknowledge that climate change has general effects on health. Only four countries (Algeria, Barbados, Iceland, Oman) do not mention general climate change related health risks at all in their national climate change documents. As Figure 5.2 illustrates, in addition to general effects of climate change on health, a large number of states (177) associates an increase of infectious diseases with climate change. The third most recognized category constitutes health risks that are connected to water and sanitation, such as waterborne diseases: 171 states acknowledge that climate change has a negative effect on such health issues. Tied on place four are heat-related health risks and health risks related to floods and storms. Food and agriculture related health risks are recognized by 139 countries and risks related to droughts and fires by 136 countries. Effects of climate change on health risks due to reduced air quality are recognized by 113 states. Only 63 states recognize that climate change may also lead to more cold-related health risks. More diffuse climate change related health risks, such as those associated with UV radiation (45), allergies (69), mental health

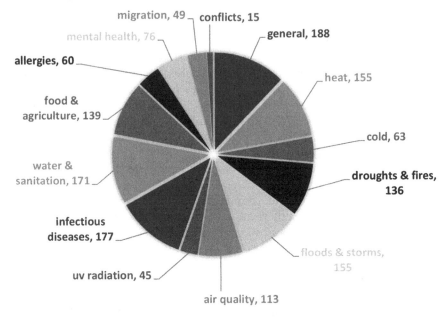

Figure 5.2 Overview of recognized health risks

(76), or migration (49), are recognized by only a small number of states. The least frequently recognized climate change related health risks are those related to conflicts, with only 15 states recognizing the nexus between climate change, conflicts, and health.

Figure 5.2 shows that the understanding of how climate change affects health differs from country to country. Generally, primary climate change related health risks are more often recognized than secondary and tertiary risks. However, despite belonging to the category of secondary risks, the effects of climate change on the spread of infectious diseases are acknowledged in a very high number of states. Moreover, climate change is less frequently associated with cold-related diseases than with heat-related health risks. Risks related to increased exposure to UV radiation or allergies receive significantly less attention than other secondary risks and, except for mental health disorders, tertiary risks are only recognized by a small fraction of the states.

Groundwork

Similar to the recognition scores, the groundwork initiatives of the countries represented in the CHAIn vary greatly. With the score of 56.25, the Kingdom of Jordan accounts for the by far highest groundwork ranking of all 192 index countries. Jordan is followed by the UK (42), North Macedonia (31.5), the Solomon Islands (30.75), Germany (27.5), Canada (23.5), the

Seychelles (22.25), the Republic of Moldova (20.25), and the Federal States of Micronesia (19.25).

At the same time, some countries that perform well in terms of groundwork initiatives show rather low adaptation scores. Jordan, for instance, has an adaptation score of 8, despite the highest groundwork score of all countries on the index. North Macedonia is an even more extreme example. Despite a very high groundwork score of 31.5, the country does not account for any adaptation initiatives whatsoever. On the other side of the spectrum, some countries with a very high adaptation score show a comparatively low performance in terms of groundwork initiatives. The state with the highest adaptation ranking on the index, the RoK, for instance, has a groundwork score of 9.75, which is significantly lower than many high performing countries on the adaptation index. Similarly, Cyprus (groundwork score of 7.5), the US (4), Turkey (8), and Nepal (3) all have low groundwork scores despite comparatively high adaptation scores.

Overall, countries reported 3,688 groundwork initiatives. As Figure 5.3 indicates, the by far highest amount of groundwork initiatives is constituted by general measures on health effects of climate change (1,434, which equals 39 percent). In terms of risk-specific initiatives, overall countries reported

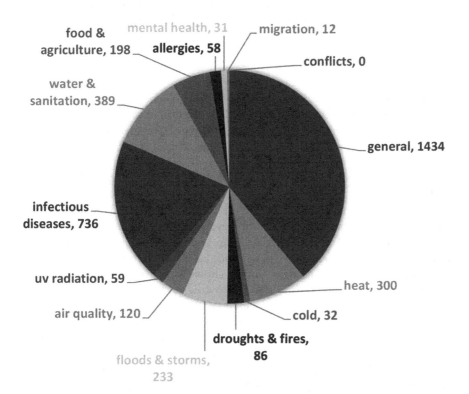

Figure 5.3 Overview of groundwork-level initiatives per health risk

736 groundwork measures that address infectious diseases (20 percent), 389 initiatives on water and sanitation related health risks (11 percent), 300 measures to counter heat-related health risks (8 percent), 233 groundwork initiatives in the category floods and storms (6 percent), and 198 measures in the field of food and agriculture (5 percent). With 120 counted initiatives, groundwork measures on reducing the effects of climate change on health risks associated with decreased air quality represent only a small share of the portfolio. Groundwork initiatives in the categories droughts and fires (86), UV radiation (59), and allergies (58) play an even less prominent role. The least addressed among secondary climate change related health risks are cold-related risks with only 32 initiatives across all states. Tertiary climate change related health risks are almost completely ignored with only 31 groundwork initiatives in the category mental health, 12 in the category migration, and no groundwork initiatives at all in the category conflicts.

The comparison of groundwork initiatives shows that most countries start with general measures on climate change and health and then focus on a small number of specific health risks, such as infectious diseases, risks related to water and sanitation, and heat-related health risks. While the rationale behind the decisions to focus on general measures as well as a small number of specific risks will be analyzed in greater detail in Chapter 6, the distribution clearly shows that cold-related health risks play a very minor role and tertiary risks are almost completely ignored.

Adaptation

The performances of the analyzed states on the adaptation index also vary to a great extent. While the RoK has an adaptation score of 88, which indicates that 44 different tangible actions have been implemented already, a great number of states has not implemented any adaptation actions at all. This includes numerous countries that perform well in terms of GDP, GDP per Capita, or other performance indexes, such as Iceland, Qatar, Luxembourg, and many others. Moreover, a significant number of developing countries shows a very strong performance with regards to their adaptation scores. Moldova (adaptation score of 24), the Seychelles (20), Sri Lanka (28), Uganda (22), and Nepal (26) all have higher adaptation scores than most developed countries, including Germany (14), France (10), and Denmark (12).

Additionally, the adaptation index shows that 89 countries have not implemented any adaptation actions at all, and only 63 countries can demonstrate three or more implemented actions. The top performing countries in terms of implemented adaptation actions are: 1. the RoK; 2. the UK; 3. Cyprus; 4. Canada; 5. the United States; 6. Sri Lanka; 7. Turkey; 8. Nepal; 9. the Republic of Moldova; and 10. Botswana.

In total, 583 health adaptation initiatives were reported, out of which 169 initiatives (29 percent) represented general actions. As Figure 5.4 shows, 149 initiatives, adaptation actions that address infectious diseases, represent 26

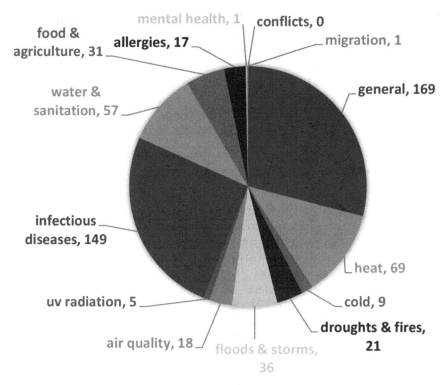

mental health, 1 · conflicts, 0

food & agriculture, 31 · allergies, 17 · migration, 1

water & sanitation, 57 · general, 169

infectious diseases, 149 · heat, 69

uv radiation, 5 · cold, 9

air quality, 18 · droughts & fires, 21

floods & storms, 36

Figure 5.4 Overview of adaptation-level initiatives per health risk

percent of all counted measures. Heat-related health risks follow with 69 counted initiatives (12 percent) and measures in the category water and sanitation were identified 57 times (10 percent). The categories floods and storms (36 initiatives, which equals 6 percent) and food and agriculture (31 initiatives, which equals 5 percent) accounted for much smaller shares in the overall adaptation portfolio. Moreover, states reported 21 adaptation initiatives (4 percent) in the category droughts and fires, 18 (3 percent) in the category air quality, and 17 (3 percent) in the category allergies. Almost no adaptation initiatives on tertiary climate change related health risks exist. Only one measure on mental health risks and one measure in the category migration were counted. No country reported any adaptation initiatives in the category conflicts.

Similar to the health risks that were addressed by states in their ground-work initiatives, their adaptation actions mostly include general measures as well health adaptation actions that focus on infectious diseases. As a matter of fact, the share of measures that address infectious diseases is even higher with regards to adaptation (26 percent) than with regards to groundwork initiatives (20 percent). Moreover, the general tendency of the health risks addressed in the groundwork initiatives, with large shares in the categories heat and water and sanitation, holds true for the adaptation actions as well. Compared with the groundwork initiatives, heat related health risks play an even more prominent role in the adaptation category since they constitute 12 percent of all adaptation initiatives (and only 8 percent of groundwork initiatives). Water and sanitation constitute 10 percent of all adaptation initiatives (versus 11 percent of groundwork initiatives). Similar to groundwork initiatives, tertiary climate change related health risks are almost completely excluded from current adaptation actions.

Adaptation types

The comparison of adaptation types (Figure 5.5) illustrates that the vast majority of reported health adaptation initiatives represent recommendations (2,514 out of 4,271 reported initiatives). The second most reported category is "strategy & plans" with 659 initiatives, followed by "research and development" with 488 initiatives. Overall, more than two thirds of all adaptation initiatives constitute groundwork-level measures. Among the comparatively few adaptation-level initiatives, most of them represent capacity building measures (169). They are followed by measures from the categories "education & public awareness" (108), "monitoring & warning" (108), and "practice & behavior" (89). In total, only 27 infrastructure measures, and 13 financing initiatives, and 11 technological innovations to adapt to climate change related health risks were reported.

The clear dominance of recommendations over all other adaptation types shows that much more work needs to be done to adapt against climate

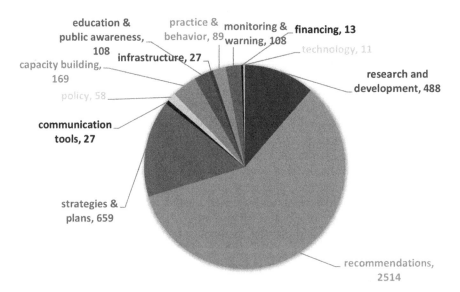

Figure 5.5 Overview of adaptation types

change related health risks, because recommendations do not imply any implemented action, but rather display what states announce they should do. Moreover, the low number of adaptation-level initiatives demonstrates how important it is to weight the factors as described above so that the index reflects which states have already implemented tangible actions to adapt against climate change related health risks and which states focus more on preparing for further action in the future.

Limitations and strengths of the index

Despite the many advantages of the index, since it is the first comprehensive overview in the field it also comes with a number of limitations that offer opportunities for further research projects to expand on the results and add new perspectives. First and foremost, the database relies on official national documents. Therefore, the quality of the index depends on how well the states report their adaptation initiatives. As the index measures and compares what states include in their official documents, a variance between their actual adaptation measures and their reported measures may exist. To complement the database and check for adaptation measures that were not included in the official documents, I initially developed a survey that was intended to be sent to all WHO country representatives of the 192 states. However, a trial run for Latin America showed a very low response rate and indicated that such an endeavor was not possible for a high number of states. Additionally, a cooperation with the World Health Organization's Department for Climate Change and Health was initiated and the WHO agreed to grant me access to additional data. However, it showed that the data of the WHO Climate and Health Country Profile Project was neither sufficient for the requirements of the CHAIn, nor did it provide enough material for all 192 countries.[11]

Additionally, the publication dates of the respective national documents vary greatly, with the most recent publications that were taken into account being published in 2018 and the oldest ones in the early 2000s. Therefore, an additional structured online search was undertaken to identify new documents and to check whether all relevant documents were identified. Moreover, the case studies helped to test whether the data generation worked effectively since, in most cases, the experts confirmed that all relevant documents had been identified. In the rare instances where they mentioned additional sources, the incorporation of those documents did not lead to any significant changes in the country rating since they did not include many additional initiatives.

Furthermore, the index focuses on national-level action. However, as the case studies show, numerous actions take place on the regional or local level and are often not included in national adaptation plans. As a consequence, even if entire countries have a high CHAIn score, that does not necessarily mean that this is the case for specific regions or cities, and the other way around. The city of New York, for instance, has undertaken a great number of health adaptation initiatives that are not included in the national adaptation strategies of the United States and are therefore not measured (Araos et al. 2016, Rosenzweig et al. 2007, Rosenzweig and Solecki 2010). Due to the high number of states and documents that had to be analyzed to construct this index, it was not possible to zoom further in and take regional and local initiatives into account.

Moreover, the index scores need to be seen in perspective. Although high-ranking countries, such as the UK or South Korea, have a much higher score than many other states, the score does not indicate whether the countries are well prepared for climate change related health risks or not.[12] Despite their comparatively high number of adaptation measures, there is still a lot of room for additional measures. More importantly, while the index seeks to track current health adaptation measures and therefore facilitate comparisons among states, it does not make statements about the quality of health adaptation to climate change of specific countries. Such assessments need to be made on a case by case level due to the large geographical, political, social, and economic differences among states.

The CHAIn index constitutes the first global assessment of health adaptation initiatives to climate change of a significantly high number of states (192). Therefore, for the first time ever, it is possible to compare how almost all UN Member States adapt to climate change related health risks, and to analyze whether the countries are prepared for the risks they are facing, what kind of actions they take, and how well they communicate their adaptation measures. With its variety of different categories, the index grants deep insights into the health risks states associate with climate change, their levels of adaptation, and the specific types of adaptation they undertake.

In addition to its significant importance for research on health adaptation to climate change and the numerous possibilities it offers for future research projects in the field, the index is of high societal relevance since it grants decision makers and the general public the opportunity to understand what needs to be done to better prepare for climate change related health risks and how countries can learn from each other.

The following chapters will delve into the reasons behind the varying performance of states on the CHAIn and its respective sub-indexes, such as the recognition, groundwork, and adaptation index. Following the mixed methods design of this study, the book will continue with an overview of potential theoretical explanations for the varying performance of states in terms of their health adaptation policies. The theoretical assumptions will then be tested by a multiple regression analysis. Following the first quantitative chapter, five case studies will contribute to a better understanding of the drivers of and barriers to health adaptation to climate change and critically reflect on the composition of the index to further improve its robustness.

Notes

1 More information on the case studies follows in Part III.
2 For further information see the codebook in the supplementary material.
3 When the initial version of the index was first presented to fellow PhD Candidates and lecturers at the University of Heidelberg, the question came up, whether the category "denial" should be established as a counterpart to recognition. After careful evaluation of both the theoretical and empirical discourse on climate change and health, it became clear that such a category does not have empirical foundation since climate change and especially the nexus between climate change and health has not been denied by any country in their official national documents that build the database of this research project.
4 For more information on how and when specific initiatives were sorted into the respective categories see the supplementary material in the supplementary material.
5 Compared with recommendations, strategies and plans are more likely to be implemented within a given time and will therefore be weighted with a higher factor as will be explained below.
6 For more information on the weighting of the factors and the different steps of the analysis, see the codebook in the Appendix.
7 As research on health adaptation to climate change is quickly evolving and state responses change frequently, the index can and should be continuously improved. The regression analyses have shown some avenues for further development on the subject, but considering that the CHAIn represents the first global health adaptation index of this kind, it certainly serves the purpose of providing a global overview and helping to understand what drives and holds back health adaptation to climate change on a global level.
8 For a full list of all analyzed documents see the codebook in the supplementary material.
9 To further enhance the dataset, I originally planned to do a survey with WHO country representatives and representatives of all states. However, an initial trial run for Latin America proved that the method was unsuccessful. After numerous reminders, not a single contact person responded to the e-mails. For further information see the interview questions in the Appendix.
10 A correlation analysis of the publication date and the CHAIn score has shown that no relation between the two factors exists. Accordingly, states with newer documents did not rank higher. Further information can be found in the do-file in the Appendix.
11 At the time of finishing this thesis, the WHO Climate and Health Country Profile was only available for a limited number of countries and even with 101 countries represented in the most recent report, the number of cases is significantly lower than in the CHAIn (191). Moreover, the WHO Climate and Health Country Profile does not provide in-depth data on the specific measures that states have undertaken based on their national reports, but rather answers a number of basic questions, such as whether the states have conducted a vulnerability assessment or whether they have health adaptation plans (WHO/UN 2019).
12 The supranational level, especially in terms of the European Union, plays a significant role in driving national-level adaptation level, as the case studies on Ireland and the UK show.

References

Araos, Malcolm, Stephanie E. Austin, Lea Berrang-Ford, and James D. Ford. 2016. "Public health adaptation to climate change in large cities: a global baseline." *International Journal of Health Services* 46 (1): 53–78.

Austin, Stephanie E., Robbert Biesbroek, Lea Berrang-Ford, James D. Ford, Stephen Parker, and Manon D. Fleury. 2016. "Public health adaptation to climate change in OECD countries." *International Journal of Environmental Research and Public Health* 13 (889): 1–20. doi:10.3390/ijerph13090889.

Biagini, Bonizella, Rosina Bierbaum, Missy Stults, Saliha Dobardzic, and Shannon M. McNeeley. 2014. "A typology of adaptation actions: a global look at climate adaptation actions financed through the Global Environment Facility." *Global Environmental Change* 25:97–108. doi:10.1016/j.gloenvcha.2014.01.003.

Biesbroek, G. Robbert, Rob J. Swart, Timothy R. Carter, Caroline Cowan, Thomas Henrichs, Hanna Mela, Michael D. Morecroft, and Daniela Rey. 2010. "Europe adapts to climate change: comparing National Adaptation Strategies." *Global Environmental Change* 20 (3): 440–450. doi:10.1016/j.gloenvcha.2010.03.005.

Butler, Colin, Devin Bowles, Lachlan McIver, and Lisa Page. 2014. "Mental health, cognition and the challenge of climate change." In *Climate Change and Global Health*, edited by Colin Butler, 251–259. Canberra: CABI.

Carter, Timothy, Martin Parry, H. Harasawa, and S. Nishioka. 1994. "*IPCC technical guidelines for assessing climate change impacts and adaptations.*" In Part of the IPCC Special Report to the First Session of the Conference of the Parties to the UN Framework Convention on Climate Change, Intergovernmental Panel on Climate Change. Department of Geography, University College London, UK and Center for Global Environmental Research, National Institute for Environmental Studies, Tsukuba, Japan.

Hanna, Elizabeth, and Lachlan McIver. 2014. "Small island states – canaries in the coal mine of climate change and health." In *Climate Change and Global Health*, edited by Colin Butler, 181–192. Canberra: CABI.

Lesnikowski, A. C., J. D. Ford, L. Berrang-Ford, M. Barrera, P. Berry, J. Henderson, and S. J. Heymann. 2013. "National-level factors affecting planned, public adaptation to health impacts of climate change." *Global Environmental Change* 23 (5): 1153–1163. doi:10.1016/j.gloenvcha.2013.04.008.

McMichael, Anthony J. 2013. "Globalization, climate change, and human health." *The New England Journal of Medicine* 368 (14): 1335.

Neira, Maria, Roberto Bertollini, Diarmid Campbell-Lendrum, and David L. Heymann. 2008. "The year 2008 – A breakthrough year for health protection from climate change?" *American Journal of Preventive Medicine* 35 (5): 424–425.

Smit, Barry, and Mark W. Skinner. 2002. "Adaptation options in agriculture to climate change: a typology." *Mitigation and Adaptation Strategies for Global Change* 7 (1): 85–114.

Smit, Barry, Ian Burton, Richard J. T. Klein, and Johanna Wandel. 2000. "An anatomy of adaptation to climate change and variability." In *Societal Adaptation to Climate Variability and Change*, edited by Gary Yohe and Sally M. Kane, 223–251. Dordrecht: Springer.

Rosenzweig, Cynthia, David C. Major, Kate Demong, Christina Stanton, Radley Horton, and Melissa Stults. 2007. "Managing climate change risks in New York City's water system: assessment and adaptation planning." *Mitigation and Adaptation Strategies for Global Change* 12 (8): 1391–1409.

Rosenzweig, Cynthia, and William Solecki. 2010. "Climate change adaptation in New York City." *Annals of the NY Academy of Science* 1196.

Tompkins, Emma L., W. Neil Adger, Emily Boyd, Sophie Nicholson-Cole, Keith Weatherhead, and Nigel Arnell. 2010. "Observed adaptation to climate change:

UK evidence of transition to a well-adapting society." *Global Environmental Change* 20 (4): 627–635. doi:10.1016/j.gloenvcha.2010.05.001.

Wilbanks, Thomas J., and Robert W. Kates. 1999. "Global change in local places: how scale matters." *Climatic Change* 43 (3): 601–628.

World Health Organization, United Nations (WHO/UN). 2019. *2018 WHO Health and Climate Change Survey Report: Tracking Global Progress*. Geneva: United Nations World Health Organization.

World Health Organization, United Nations, and United Nations World Meteorological Organization. (WHO/WMO). 2012. *Atlas of Health and Climate*. Geneva: World Health Organization.

6 Regression results

Based on the theoretical and empirical evidence on potential drivers of and barriers to health adaptation to climate change, national-level predictors were selected to test the following hypotheses:

H1: The more vulnerable a state is to climate change related health risks, the more it does to adapt to climate change related health risks.

H2: The better the economic health (measured in GDP PPP) of a state is, the more it does to adapt to climate change related health risks.

H3: The higher the population size of a state is, the more it does to adapt to climate change related health risks.

H4: The more a state guarantees civil liberties, political rights, and regularly held, free and fair elections, the more it does to adapt to climate change related health risks.

H5: The lower the perceived corruption of a state is, the more it does to adapt to climate change related health risks.

H6: The stronger the influence of international organizations is on the state, the more it does to adapt to climate change related health risks.

H7: The stronger the epistemic community within a state is, the more it does to adapt to climate change related health risks.

H8: The younger the population of a state is, the more the state does to adapt to climate change related health risks.

In the following, the data, methods, and results will be presented and discussed, and the cases for the SNA will be selected based on the results of the LNA.

Data

The dependent variables that will be tested in several different constellations are based on the CHAIn. First, several national-level predictors for the overall CHAIn will be analyzed. Afterwards, the same predictors will be utilized to analyze their influence on the recognition-level index, the groundwork-level index, and the adaptation-level index. Table 6.1 summarizes the predictors, indexes, and sources.

Table 6.1 Dataset of the first multiple regression analysis

Variable	Index	Year	Source	Countries
Climate change vulnerability	ND-GAIN vulnerability	2017 (data from 2016)	Notre Dame University	181
Economic well-being	GDP, PPP 2016 (current international $)	2016	World Bank	230
Degree of Civil Rights and Political Freedom	Freedom in the World	2018 (data from 2017)	Freedom House	210
Good governance	Corruption Perceptions Index 2017	2017	Transparency International	180
Influence of international organizations (dummy variable)	Documents that were analyzed for the CHAIn	2018	CHAIn database	192
Epistemic communities	Research and development expenditure (% of GDP)[1]	2016 (if not available, earlier)	World Bank	121
Generational effects	Median age	2019	CIA World Factbook	191

The vulnerability of states to climate change will be measured by the ND-GAIN index of the Notre Dame University in Notre Dame, Indiana, United States. The index data is from 2016 and includes various areas of vulnerability, such as food, water, health, infrastructure, and others (ND-GAIN 2019). Although Notre Dame University provides a specific index for health-related climate change vulnerabilities, the general ND GAIN Vulnerability Index was used because it is more comprehensive and includes more climate change related health risks than the health-specific one.[2] As the literature on health adaptation to climate change frequently discusses economic variables as potential predictors of health adaptation, economic well-being (GDP (PPP)) of countries will be tested.[3]

The "Degree of Civil Rights and Political Freedom," measured by Freedom House's Freedom in the World Index 2018 (data from 2017), indicates the level of civil rights and political freedom in the respective countries. Additionally, following Berrang-Ford et al. (2014), the overall level of good governance may serve as a predictor for health adaptation. Therefore, Transparency International's CPI with data from 2017 has been included in the analyses as a proxy for good governance.

The influence of international organizations on the national-level strategies and plans for health adaptation to climate change was operationalized by a dummy variable, specifically created for this study, that distinguishes whether international organizations, such as UN Environment, the UNDP, the WHO, the UNFCCC, the World Bank, and others, financed or in any other way supported the development of NCs, National Adaptation Plans, and other documents that summarize states' health adaptation measures to climate change. If international organizations contributed to the documents, the country was encoded as "1," if not, it received a "0."

To measure the strength of epistemic communities, the World Bank's data on research and development expenditure as percentage of the GDP was added to the data set as a proxy. If the variable did not provide sufficient information on a country's research and development expenditure in 2016 or later years, data from earlier years was added if it was provided by the World Bank. Other variables, such as the number of PhDs in the respective countries, were also considered as proxies for epistemic communities. However, not enough data was available to meet the needs of the large dataset.

Methods and tests

To test the influence of the independent variables on the dependent variable, a multiple regression analysis, using STATA (StataCorp v.13), was conducted. Relationships were considered significant at the 95 percent confidence level. Information on regression diagnostics can be found in the supplementary material. Numerous tests, including tests for multicollinearity, heteroscedasticity, the influence of residuals etc. were conducted. Although the first four regressions showed higher explanatory powers (R-Squared) than the initial regressions before the case studies and second theoretical chapter, they were not robust and satisfactory. Moreover, significances changed for all dependent variables, except for the recognition-level index, which was still significantly influenced by the variance on the Freedom House index. As the case studies show, many countries have developed a high number of groundwork-level initiatives, such as recommendations or strategies, which can ultimately lead to higher CHAIn scores. Although groundwork-level measures were weighted accordingly in the index, some states had such a high amount of recommendations or strategies and plans in their documents that they influenced the general trends and distorted the analysis. Therefore, some of the cases were taken out of the analysis and all four regressions were run again. As Pollet and van der Meij (2017) suggest, it is possible to remove outliers and compare the results with the initial regression results if everything is documented transparently. The entire documentation can be found in the do-file in the Appendix. The new regressions were much more robust and satisfactory and will be presented below.

Results

The four regressions led to numerous new findings on the drivers of and barriers to health adaptation to climate change, which will be presented in the following. Further information can be found in the do-file in the Appendix.

CHAIn

The regression analysis that measures the influence of the predictors on the variance of the CHAIn has led to much more robust findings than all previous regressions (see Table 6.2). With an R-Squared of 0.401 in the best-fit model (model 5), 40 percent of the variation on the CHAIn can be explained by the variation of the independent variables. With 92 cases, the sample is much larger than for all existing studies on adaptation to climate change.

The regression analysis that measures the influence of the predictors on the variance of the CHAIn has led to much more robust findings than all previous regressions (see Table 6.2). With an R-Squared of 0.401 in the best-fit model (model 5), 40 percent of the variation on the CHAIn can be explained by the variation of the independent variables. With 92 cases, the sample is much larger than for all existing studies on adaptation to climate change.

Table 6.2 CHAIn models 2. Multiple regression analysis

	Model 1 b/se	Model 2 b/se	Model 3 b/se	Model 4 b/se	Model 5 b/se
GDP (PPP)	0.000***	0.000***	0.000***	0.000***	0.000***
	(0.00)	(0.00)	(0.00)	(0.00)	(0.00)
Vulnerability (ND-~)		−5.885	−12.646	−10.669	−10.973
		(8.84)	(10.15)	(12.98)	(19.08)
Freedom House		0.082**	0.082**	0.037	0.041
		(0.03)	(0.03)	(0.03)	(0.04)
International orga~s			2.042	6.131**	5.969**
			(1.85)	(1.96)	(2.09)
Research and devel~t				6.494***	6.947***
				(1.44)	(1.80)
Median age (CIA Wo~2					0.001
					(0.20)
CPI					−0.033
					(0.08)
Constant	13.392***	11.292*	13.025*	9.052	10.138
	(0.74)	(5.00)	(5.06)	(6.42)	(13.48)
R-Sqr	0.104	0.176	0.185	0.400	0.401
dfres	165	155	153	94	92
BIC	1227.2	1168.7	1162.4	722.3	731.3

Note: * $p<0.05$, ** $p<0.01$, *** $p<0.001$.

The regression indicates that a strong relationship between the independent variable "GDP" and the dependent variable "CHAIn" exists until the introduction of the independent variable "epistemic communities," for which the proxy expenditure of research and development as percentage of the GDP was utilized. Similarly, the independent variable "Freedom House" loses significance with the introduction of additional independent and control variables in models 4 and 5 and is no longer significant in the best-fit mode. The independent variable "epistemic communities" is highly significant in models 4 and 5 and can therefore be considered as the major predictor for the CHAIn's variance. The proxy for the influence of international organizations also shows significant influence on the dependent variable ($p<0.01$). The CPI as a proxy for good governance and the median age of the population as a proxy for generational effects do not show any significant effects.

Recognition

The regression results with the recognition-level index as dependent variable are very similar across all calculations and gain further explanatory power with the most recent analysis (see Table 6.3). With an R-Squared of 0.247 in

Table 6.3 Recognition-level index models 2. Multiple regression analysis

	Model 1 b/se	Model 2 b/se	Model 3 b/se	Model 4 b/se	Model 5 b/se
GDP (PPP)	0.000 (0.00)	0.000 (0.00)	0.000 (0.00)	−0.000 (0.00)	−0.000 (0.00)
Vulnerability (ND-~)		−2.715 (2.83)	−2.823 (3.31)	−5.127 (4.68)	−0.991 (6.85)
Freedom House		0.037*** (0.01)	0.037** (0.01)	0.035** (0.01)	0.031* (0.01)
International orga~s			0.026 (0.60)	0.299 (0.71)	0.520 (0.75)
Research and devel~t				0.738 (0.52)	0.538 (0.65)
Median age CIA Wo~2					0.056 (0.07)
CPI					0.011 (0.03)
Constant	7.864*** (0.25)	6.917*** (1.60)	6.952*** (1.65)	7.426** (2.31)	3.623 (4.84)
R-Sqr	0.000	0.144	0.143	0.240	0.247
dfres	165	155	153	94	92
BIC	855.9	806.9	808.0	518.2	526.4

Note: * $p<0.05$, ** $p<0.01$, *** $p<0.001$.

the best-fit model (model 5), almost 25 percent of the variation on the recognition-level index can be explained by the variation of the independent variables.

Across all models, the regression shows that states with higher Freedom House ratings tend to have higher recognition levels. Although the significance decreases from $p<0.001$ in model 2 to $p<0.05$ in the best-fit model (model 5), the predictor is significant throughout all regressions that were calculated as part of this thesis and can therefore be considered satisfactory.

Groundwork

The regressions with the groundwork-level index as dependent variable were the least robust across all regressions. With the most recent regression analysis, the explanatory power was enhanced to R-Squared of 0.165 and, for the first time, predictors proved to have a significant relationship with the dependent variable (see Table 6.4).

Until the introduction of the independent variables "influence of international organizations" and "epistemic communities," the regression shows a significant relationship between the independent variable "Freedom

Table 6.4 Groundwork-level index models 2. Multiple regression analysis

	Model 1 b/se	Model 2 b/se	Model 3 b/se	Model 4 b/se	Model 5 b/se
GDP (PPP)	0.000 (0.00)	0.000 (0.00)	0.000 (0.00)	0.000 (0.00)	0.000 (0.00)
Vulner-ability (ND-~)		4.766 (4.92)	0.196 (5.69)	-8.142 (7.79)	5.567 (11.12)
Freedom House		0.037* (0.02)	0.037* (0.02)	0.010 (0.02)	0.012 (0.02)
Interna-tional orga~s			1.578 (1.04)	2.791* (1.18)	2.973* (1.22)
Research and devel~t				1.371 (0.86)	2.530* (1.05)
Median age CIA Wo~2					0.205 (0.11)
CPI					-0.084 (0.04)
Constant	5.514*** (0.40)	1.335 (2.78)	2.303 (2.84)	6.098 (3.85)	-3.452 (7.85)
R-Sqr	0.014	0.046	0.060	0.111	0.165
dfres	165	155	153	94	92
BIC	1022.5	982.6	979.7	620.2	623.2

Note: * $p<0.05$, ** $p<0.01$, *** $p<0.001$.

House" and the dependent variable "groundwork-level index" (models 2 and 3). In the best-fit model (model 5), the independent variables "research and development" and "international organizations" are significant at p<0.05.

Adaptation

The robustness of the regression analysis that centers around the dependent variable "adaptation-level index" significantly improved with the latest version of the regression. R-Squared is now at 0.391, which means that almost 40 percent of the variation on the adaptation-level index can be explained by the variation of the independent variables.

With the introduction of additional independent variables in models 4 and 5, the significance between the independent variable GDP (PPP) reduces. Although the best-fit model (model 5) shows a statistically significant effect between GDP and adaptation-level measures, the effect is less significant than in all previous regressions (p<0.01). At the same time, epistemic communities, measured by expenditure on research and development, show a very high significance (p<0.001) in model 4 and high significance in model 5 (p<0.01).

Table 6.5 Adaptation-level index models 2. Multiple regression analysis

	Model 1 b/se	Model 2 b/se	Model 3 b/se	Model 4 b/se	Model 5 b/se
GDP (PPP)	0.000*** (0.00)	0.000*** (0.00)	0.000*** (0.00)	0.000** (0.00)	0.000** (0.00)
Vulnerability (ND-~)		−9.293 (5.96)	−11.430 (6.84)	0.037 (9.57)	−16.044 (13.85)
Freedom House		0.030 (0.02)	0.027 (0.02)	0.009 (0.03)	0.014 (0.03)
International orga~s			0.452 (1.24)	3.190* (1.45)	2.736 (1.15)
Research and devel~t				4.754*** (1.06)	4.125** (1.31)
Median age CIA Wo~2					−0.233 (0.14)
CPI					0.046 (0.06)
Constant	3.945*** (0.49)	6.498 (3.37)	7.235* (3.41)	−0.760 (4.73)	11.778 (9.78)
R-Sqr	0.145	0.188	0.196	0.371	0.391
Dfres	165	155	153	94	92
BIC	1088.7	1043.5	1037.7	661.3	667.2

Note: * p<0.05, ** p<0.01, *** p<0.001.

Regression diagnostics

A great number of tests has been conducted to evaluate whether the regression results show signs of multicollinearity or heteroscedasticity, and whether the residuals influence the regression results. The mean variance inflation factor for the best-fit model with the CHAIn as dependent variable is 2.67 and thus not problematic. After performing the White's general test, the null hypothesis, which states that homoscedasticity exists, cannot be rejected since the p-value is 0.3505.[4] The graphical analysis of the residuals of the independent variable did not show a clear trend concerning the residuals and the dependent variables. Therefore, all regressions were calculated with robust standard errors. The results were similar, but for the CHAIn and the adaptation-level index as dependent variables, GDP had a more significant impact on them than in the previous regression analysis. To further investigate the effect of the residuals on the regression, Cook's D was calculated. According to Cook's D, Greece, Guatemala, Lithuania, Mali, and China were influential cases. To test the strength of their index, they were excluded from the analysis and the regression was run again. The analysis showed that the models only slightly differ in their explanatory power with the R-Squared decreasing to 0.3982 for the CHAIn as dependent variable. Moreover, the effect of GDP and International Organizations on the CHAIn becomes slightly stronger. The effect of epistemic communities remains very strong.

Interpretation of results

The regressions have partially confirmed the theoretical assumptions, yet the subject requires further studies to enhance the explanatory power of the dependent variables and to test other potential independent variables. Importantly, the Freedom House score significantly influences the recognition-levels of climate change related health risks across all models, which suggests that the more states guarantee political rights, civil liberties, and regularly held free and fair elections, the more climate change related health risks they recognize (hypothesis 2). Furthermore, epistemic communities and international organizations have a strong effect on how states rank on the CHAIn and especially on how many groundwork-level initiatives they report on. Adaptation-level initiatives are heavily influenced by epistemic communities and GDP. Although GDP remains a significant predictor, its influence is less significant than in previous regressions. As this is the first global study that systematically compares such a high number of states, further proxies and variables need to be tested. However, the regression results largely confirm the findings from the case studies and thus show that the Sieve Model on health adaptation to climate change makes an important and valid contribution to research on this topic.

To summarize, the multiple regression analysis confirmed the explanatory model. Epistemic communities and international organizations are among the key drivers of health adaptation to climate change since they spread information on the topic, frame the discourse, support governments in developing strategies and plans, and are sometimes involved in adaptation-level projects. Due to the specific characteristics of democracies compared to autocracies, such as their openness towards new and diverse ideas, recognition levels increase with rising Freedom House ratings. The economic health of a country is an important factor as well, especially when it comes to adaptation-level measures. Additionally, international organizations are a key driver for states' health adaptation initiatives, especially with regards to their groundwork-level measures.

Notes

1 While other proxies, such as the Educational attainment, Doctoral or equivalent, population 25+, total (%) (cumulative), World Bank, 2017 or newer, were considered, the research and development expenditure in % of the GDP provided the largest database. Data on other proxies was not available for a large number of states. In future studies, it will definitely be worth analyzing additional proxies that capture the strength of epistemic communities.
2 Since the index has certain disadvantages, an additional dummy variable based on Bhatt et al. (2013)'s prominent article on the spread of dengue in the world has been created and added to the analysis.
3 In the process of building the model, additional socioeconomic variables such as GDP per Capita and the United Nations Population Division's Population Index were tested. However, they did not show any significant impact on the CHAIn or its sub-indexes. Further details can be found in the supplementary material.
4 If the p-value is very low (e.g. below 0.05), the null hypothesis can be rejected with a certain significance level and heteroscedasticity exists.

References

Berrang-Ford, Lea, James D. Ford, Alexandra Lesnikowski, Carolyn Poutiainen, Magda Barrera, and S. Jody Heymann. 2014. "What drives national adaptation? A global assessment." *Climatic Change* 124: 441–450.
Bhatt, Samir, Peter W. Gething, Oliver J. Brady, Jane P. Messina, Andrew W. Farlow, Catherine L. Moyes, John M. Drake, John S. Brownstein, Anne G. Hoen, Osman Sankoh, Monica F. Myers, Dylan B. George, Thomas Jaenisch, G. R. William Wint, Cameron P. Simmons, Thomas W. Scott, Jeremy J. Farrar, and Simon I. Hay. 2013. "The global distribution and burden of dengue." *Nature* 496 (7446): 504–507. doi:10.1038/nature12060.
ND-GAIN. 2019. "Country index." Accessed 23 May 2019. https://gain.nd.edu/our-work/country-index/.
Pollet, Thomas V., and Leander van der Meij. 2017. "To remove or not to remove: the impact of outlier handling on significance testing in testosterone data." *Adaptive Human Behavior and Physiology* 3 (1): 43–60. doi:10.1007/s40750-016-0050-z.

Part III
The national perspective

7 Qualitative research design and case selection

Following the nested analysis design by Lieberman (2005, 435), the cases for the qualitative part were deliberately chosen to build a more robust model through the SNA. Based on the findings of the LNA, the following cases were selected: Japan, the RoK, the Republic of Ireland, the UK, and Sri Lanka. The RoK, the UK, and Sri Lanka ranked much higher than expected, while the Republic of Ireland and Japan ranked lower than expected. At the same time, the cases represent a wide range of different characteristics, such as in terms of the GDP (e.g. the UK has a comparatively high GDP whereas Sri Lanka's GDP is comparatively low), population size (Japan has a large population whereas Ireland has a comparatively small population), geographical distribution (two European, two East-Asian, and one South-Asian country), Freedom House rating (e.g. the UK is listed as free, whereas Sri Lanka is rated as partly free), and many more. As a consequence, the SNA helps to dig into the potential drivers of and barriers to health adaptation in the respective countries, which can subsequently be tested for general trends in the last quantitative part. The SNA connects with the LNA through the case selection and through building a new explanatory model that can subsequently be tested. Therefore, the study takes full advantage of the mixed methods design by making use of the respective advantages of qualitative and quantitative methods and balancing out their individual disadvantages.

The case studies rest on four major pillars: a) a brief summary of the major climate change related health risks for the respective state, b) an additional qualitative content analysis that applies the analytical framework of the CHAIn index to previous NCs of that state in order to identify changes over time, c) a structured literature review of the health adaptation initiatives and potential drivers of and barriers to adaptation, and d) handbook-supported expert interviews with representatives from academia, politics, and international organizations in the respective country.

Health risk assessment

The goal of the brief health risk assessments is to better understand the exposure and vulnerability of the selected countries to climate change related health risks and to analyze whether different levels of vulnerability or exposure influence state perceptions and decisions on health adaptation to climate change. Since no prominent global vulnerability index specifically for climate change related health risks exists yet, the health risk assessment had to combine several different sources, including the Notre Dame Global Adaptation Initiative Index, information from the World Bank Climate Change Knowledge Portal, and significant academic articles on the vulnerability of the respective countries to climate change related health risks. Due to lacking comprehensive and comparable data on the specific climate change related health risks countries are facing, the health risk assessment is not sufficient to provide a detailed and full-scale vulnerability assessment for each country, but rather serves as a starting point for further research that helps to develop a better understanding for the potential impact of different vulnerability levels of the cases.

Longitudinal analysis

To better understand how and why the health adaptation policies have developed over time and which role specific events played, the longitudinal analysis starts with the first NCs of the countries to the UNFCCC. To ensure comparable results, the qualitative content analysis that was used to develop the CHAIn will be applied to all documents. As a consequence, the documents will be integrated into the existing country files in MAXQDA, and graphs for the CHAIn scores over time as well as for the recognition, groundwork, and action scores over time will be created. The longitudinal analysis thus offers new insights into the perceptions of climate change related health risks in the respective states, as well as the development of their health adaptation portfolios over time.

Structured literature review

The literature review of the health adaptation policies is based on a comprehensive and structured online search of the websites of the ministries and governmental agencies in the fields of health and environment of the respective states, the University of Heidelberg's online library catalogue (HEIDI), the University of Heidelberg's online journal database that gives access to more than 120,000 e-journals, Google Scholar, and snowball sampling via prominent articles in the field. Moreover, various experts from academia and the public sector that were interviewed for this study recommended additional sources that were taken into account. The review helps to further test the reliability of the CHAIn index, since it shows whether the health adaptation initiatives that were identified for the index are complete and whether the strategies and plans are implemented on the ground level. Additionally, some articles contribute to a better understanding of the drivers of and barriers to health adaptation to climate change.

Expert interviews

The handbook-guided expert interviews with scientists, government officials, and representatives of international organizations, such as the World Health Organization or the World Bank, constitute the most essential part of the case studies since the interview partners provided exclusive information on how climate change related health risks are perceived, which actions are taken, and what drives and prevents comprehensive health adaptation initiatives to climate change in the respective countries. Many of the experts have been directly involved in the policy-making processes or have either worked with the government or put pressure on the government to take action. Additionally, numerous experts from academia were able to provide information on the effects of climate change on health in the specific countries and on the actions of the key players in the country.

Handbook-guided expert interviews constitute a powerful tool for discovering a mostly unknown research subject with high levels of uncertainty since they allow researchers to delve into the details of the topic whilst keeping the research process structured and transparent (Blatter, Janning, and Wagemann 2007, 60). This type of interview significantly differs from standardized interviews with a strict set of questions, which is more often used in completely quantitative studies because it allows the respondents to elaborate on their own perspective. Furthermore, its loose structure enables the interviewer to ask follow-up questions and make the conversation feel more natural (Blatter et al. 2007, 60). Compared with other interview types, handbook-guided expert interviews are less structured and thus grant the interviewer more flexibility, but they also guarantee that the interviewer asks a number of questions in all interviews, which ensures that the interview results can be compared with each other (Gläser and Laudel 2010, 142). As a consequence, they include both narrative elements, where the interviewed person explains the details of the research subject in more detail, and focused elements, which require shorter questions and answers (Meuser and Nagel 2009, 476).

Before the interviews were conducted, the interviewer developed a handbook to ensure that the process and results were comparable, and all relevant questions were identified beforehand. During the interviews, a standard set of questions was always asked before the experts were given the chance to add their own perspectives.[1] Within the mixed methods design of this study, the interviews followed a more inductive logic and intended to contribute to the development of new explanations of how and why states respond to climate change related health risks. Therefore, the interviews granted the interviewed experts as much flexibility as possible, whilst adhering to the handbook. The interviewees were selected after a structured online search of the key experts in the field. The major focus of the search was on authors of relevant academic literature and policy documents on the topic, governmental officials, and representatives of international organizations. Moreover, every time an expert was contacted, he or she was asked to provide additional recommendations on who should be interviewed too. In total, 32 experts were interviewed.

The cases

The five cases, the UK, Ireland, the RoK, Japan, and Sri Lanka, differ to a great extent with regards to their health adaptation initiatives to climate change as well as their socioeconomic, political, geographical, and cultural backgrounds. Moreover, they shall not be compared with each other but rather analyzed on an individual basis in greater detail in order to allow general conclusions at the end. Despite their differences, all cases are definitely worth further examination since they all represent residuals from the regression analysis that was conducted in the first quantitative part. The case studies start with the UK, the top country on the CHAIn, continue with Ireland, which ranks lower than expected, the RoK, which constitutes the second highest country on the CHAIn, and Japan, which again ranks much lower than expected. Last but not least, Sri Lanka, a very high performing country on the CHAIn will complete the case studies.

Although the case studies follow an inductive, theory-building research logic, to increase their readability and comprehensibility, they all follow the same structure. They start with a brief index test to evaluate whether the CHAIn scores represent the state of health adaptation to climate change in the respective countries or whether the index needs to be adjusted. It is followed by an overview of the health adaptation governance in the respective countries, which includes a brief international comparison, an in-depth look into the health adaptation portfolios and the institutional framework, as well as an overview of the development of the measures over time. After the summary of health adaptation governance in the respective countries, the case studies proceed with an analysis of the key drivers of and barriers to health adaptation to climate change. This includes a brief assessment of the climate change related health risks the government is confronted with to evaluate whether the government's adaptation portfolio addresses the risks the country is most affected by and whether high risk levels lead to strengthened adaptation work. The risk assessment is followed by an analysis of the risk perceptions and political agenda of the countries, their approach towards the complexity and uncertainty of the topic, social and cultural drivers of and barriers to health adaptation to climate change, economic factors, and the influence of the international community. A summary at the end of each case study ties together the different aspects and support the development of a new theoretical approach towards health adaptation to climate change.

Note

1   More information can be found in the interview handbook in the supplementary material.

References

Gläser, Jochen, and Grit Laudel. 2010. *Experteninterviews und qualitative Inhaltsanalyse als Instrumente rekonstruierender Untersuchungen*, 4th ed. Wiesbaden: Verlag Springer.

Joachim Blatter, Frank Janning, and Claudius Wagemann. 2007. *Qualitative Politikanalyse: eine Einführung in Forschungsansätze und Methoden*, 1st ed. Wiesbaden: Verlag Springer.

Lieberman, Evan S. 2005. "Nested analysis as a mixed-method strategy for comparative research." *American Political Science Review* 99 (3): 435–452. doi:10.1017/S0003055405051762.

Meuser, Michael, and Ulrich Nagel. 2009. "Das Experteninterview – konzeptionelle Grundlagen und methodische Anlage." In *Methoden der vergleichenden Politik- und Sozialwissenschaft*, edited by Susanne Pickel, Gert Pickel, Hans-Joachim Lauth and Detlef Jahn, 465–479. Wiesbaden: Verlag Springer.

8   The United Kingdom of Great Britain and Northern Ireland

Index test

The case study has shown that the UK's CHAIn score generally reflects the country's performance in the field of climate change and health and that most of the relevant documents on the UK's health adaptation to climate change had been identified as part of the analysis for the CHAIn. The documents that were considered in the UK's assessment for the CHAIn include: a) the most recent NC to the UNFCCC in the respective research period (NC 7 from 2017), the Health Protection Agency's report "Health Effects of Climate Change in the UK 2012," and the British government's "National Adaptation Programme" from 2013.

The expert interviews revealed that, after the end of the assessment period for the UK's CHAIn score, a new National Adaptation Programme had been published. Although the new document contains additional information on the UK's progress in the field of climate change and health, it cannot be added to the CHAIn database since a time freeze was in place once the assessment started in order to ensure that the results are comparable. Moreover, the interviews showed that additional key documents for health adaptation to climate change in the UK are the Climate Change Act of 2008, the UK Climate Projections (UKCP), and the Climate Change Risk Assessments.[1] The UKCP constitutes a climate analysis tool and is produced by the UK's Met Office, the meteorological agency of the UK government. It was first published in 2009 and the newest document stems from 2018. The Climate Change Risk Assessment gives an overview of all climate change related risks for the UK and was first published in 2012. It is updated every five years. In addition to the general documents on adaptation to climate change, some publications specifically for the health sector have been released in recent years. In addition to the Health Protection Agency's documents on the health effects of climate change, which were already included in the CHAIn assessment, the National Heatwave Plan from 2004 integrated the UK's adaptation policies against heatwaves. Although the information there is relevant for the country's health adaptation to climate change, the health-related aspects were later included in the national adaptation program and the health effects of climate change document and are thus captured by the CHAIn assessment.

Consequently, the vast majority of health adaptation initiatives were identified by the CHAIn assessment. The index score is thus a true representation of the overall national-level health adaptation actions to climate change that the UK reported. Moreover, the handbook-guided expert interviews and the review of relevant peer-reviewed journal articles have shown that the majority of policies and programs that the British government reported on in their official documents are actually rooted on the ground and involve concrete actions. Therefore, it can be assumed that the index score constitutes a representation of the actual level of health adaptation to climate change in the UK.

Governance

The UK's response to climate change related health risks differs greatly from that of the rest of the world. To better understand what the UK does against such risks and how it works, this section starts with an international comparison of the UK's health adaptation portfolio and continues with an overview of the institutional framework and the key actors in the field, before it moves to the discussion of the UK's main policies and changes over time.

International comparison

With an overall score of 109, the UK ranks at number 1 on the CHAIn, followed by the RoK (CHAIn score of 103.25), Jordan (71.25), Canada (70.5), and Cyprus (61). Although the RoK has implemented even more adaptation-level actions (adaptation score of 88), the UK is among the countries with the highest adaptation score (60). Moreover, it has the second highest groundwork score of all countries (42), with only Jordan having a higher groundwork score (56.25). Other countries that rank high on the CHAIn, such as the RoK, Canada, or Cyprus, have a significantly lower groundwork score and are mainly at the top due to their high adaptation score. The UK, however, has a solid balance between groundwork and adaptation, and it also has the highest possible recognition score (7).

Health adaptation portfolio

In its official national documents, the UK recognizes all health risks that academia currently associates with climate change, which shows that the British government is aware of a wide range of risks. As Figure 8.1 underlines, in terms of its groundwork-level measures, the UK has a strong tendency towards general measures (24) and initiatives that address health risks associated with floods and storms (17). Groundwork-level initiatives against

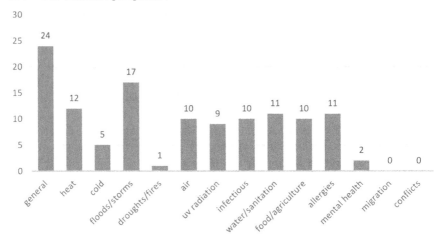

Figure 8.1 Groundwork-level initiatives in the UK

heat-related health risks rank at place 3, with 12 measures in total, whereas cold-related health risks are only addressed in five groundwork-level initiatives. With ten groundwork-level measures against climate change related health risks in relation with air pollution and nine risks related to UV radiation, the UK does much more in this field than many other states. The UK's groundwork-level initiatives on secondary health risks all range between ten and eleven measures per risk. Groundwork-level measures that address tertiary risks, such as mental health risks (2 groundwork initiatives), health risks related to migration (no measures), or health risks related to conflicts (no measures), are almost nonexistent.

Looking at Figure 8.2, it becomes clear that by far the most adaptation-level initiatives address climate change and health in general and are not risk-specific. The official documents did not mention any heat-related adaptation actions or implemented adaptation measures that address allergies or risks associated with climate change related migration or conflicts. The categories droughts and fires, UV radiation, infectious diseases, and mental health accounted for one adaptation initiative each, whereas two health adaptation initiatives on cold-related health risks were reported on. Floods and storms, the same as risks related to air pollution, water and sanitation, as well as food and agriculture, accounted for three adaptation initiatives each. Compared with groundwork initiatives, the discrepancy between general measures and risk-specific measures is even higher and many risk-specific groundwork-level initiatives, such as those that focus on heat-related risks or allergies, are not followed up with implemented adaptation initiatives. This shows that, despite some progress in terms of implemented adaptation actions, the majority of the reported health adaptation work of the UK is still on the groundwork-level and most of its implemented adaptation actions do not address concrete risks.

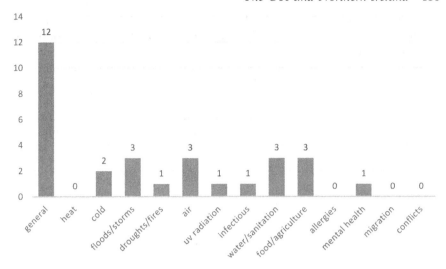

Figure 8.2 Adaptation-level initiatives in the UK

At the same time, the vast majority of adaptation measures that the UK reported on are recommendations and do not indicate that action has been taken already. Moreover, a great number of strategies and plans, which include concrete steps on what exactly should be done and until when, have been developed over the years (26 reported measures). At the same time, 19 research and development projects were mentioned in the official documents, which shows that the UK's health adaptation work has a strong research background. Two communication tools complete the groundwork portfolio of the UK, which results in 122 total groundwork-level initiatives. Compared with the 26 adaptation-level initiatives, the number of

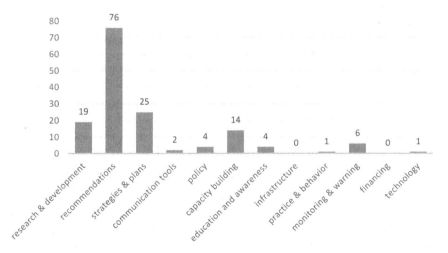

Figure 8.3 Adaptation types in the UK

groundwork-level initiatives clearly dominates the UK's health adaptation portfolio. However, it needs to be taken into account that adaptation-level actions are usually more complex and take more efforts to implement than groundwork-level actions, which is why they are also weighted differently on the CHAIn. The adaptation-level actions themselves consist mostly of capacity building measures (14), monitoring and warning systems (6), and education and awareness programs (4). One measure in the category practice and behavior and one technology instrument were reported, while no infrastructure or financing measures were mentioned in the official documents.

Institutional framework and key actors

Whilst the UK undertook some work on health adaptation to climate change already in the early 2000s, especially with regards to protection from heatwaves, the government's efforts to respond to climate change related health risks really took off with the adoption of the Climate Change Act in 2008 (Oxford University, personal communication, 27 November 2018). It led to the establishment of a great number of new institutions in the field of climate change adaptation, including the Committee on Climate Change and the Adaptation Sub-Committee of the Committee on Climate Change, and the expansion of Public Health England's Climate Change group and the health adaptation work of other governmental agencies (Oxford University, personal communication, 27 November 2018; London School of Hygiene and Tropical Medicine (LSHTM), personal communication, 25 November 2018). The Adaptation Sub-Committee monitors the government's progress on adaptation to climate change, including the National Adaptation Programme, and it requires the respective governmental branches to report to the committee on how they take action (LSHTM, personal communication, 25 November 2018).

All of the government's adaptation initiatives to climate change are coordinated and integrated by the Department for Environment, Food & Rural Affairs (DEFRA) (DEFRA, personal communication, 28 November 2018). The respective departments, such as the Department of Health and Social Care (DHSC), have the responsibility to develop their own sector-specific adaptation plans and programs and DEFRA gathers the information and integrates them in joint documents to provide a better overview and enable cooperation among the respective departments (DEFRA, personal communication, 28 November 2018). Moreover, DEFRA works closely with governmental agencies and other partners, such as Public Health England (PHE) and the Sustainable Development units in the UK (DEFRA, personal communication, 28 November 2018). One of DEFRA's major responsibilities is to produce the National Adaptation Program, which is updated every five years and was first published in 2013 (DEFRA, personal communication, 28 November 2018). It follows the release of the climate change risk assessments by the Adaptation Sub-Committee (DEFRA, personal communication, 28 November 2018). The Climate Change Risk Assessment summarizes all climate change related risks for the UK and thus helps to identify where adaptation is necessary (Oxford University, personal communication, 27 November 2018). It contains information on a great variety of risks, including health risks (Oxford University, personal communication, 27 November 2018). The Climate Change Risk Assessment is built on the results of the UK Climate Projections (UKCP), which are published by the Met Office, the UK's national weather service, and were first released in 2009 (Oxford University, personal communication, 27 November 2018). The most recent projection stems from 2018.

Whilst the overall adaptation process is coordinated by DEFRA, the Department of Health has the responsibility to develop health-related adaptation policies and programs (Oxford University, personal communication, 27 November 2018). It needs to be recognized, however, that the department's capacity to deal with the impact of climate change on health is comparatively low. According to experts, only one part-time position is dedicated to the subject programs (Oxford University, personal communication, 27 November 2018). Therefore, governmental agencies, such as Public Health England or the Sustainable Development Unit advise and support the ministry's work on a frequent basis and have a strong influence on the government's policies in the field (Oxford University, personal communication, 27 November 2018).

Public Health England is an Executive Agency to the Health Department and was established to protect and improve the health and well-being of the British people and to reduce health inequalities among them (Oxford University, personal communication, 27 November 2018). Its activities include advocacy work, networking, research, and delivering public health services (Oxford University, personal communication, 27 November 2018). Out of the 5,500 employees of Public Health England, around 900 work under the Health Protection Directorate (Oxford University, personal communication, 27 November 2018). Around 350 serve under the Centre for Radiation, Chemical and Environmental Hazards, which is again divided into the Chemicals & Environmental Effects Department and the Environmental Hazards Department. The Climate Change Group (six people) is based under the Chemicals and Environmental Effects Department and provides scientific research, advice and guidance to the UK government, local authorities, and other stakeholders (Oxford University, personal communication, 27 November 2018). It participates in international and national research projects and has strong links with universities and other research organizations (Oxford University, personal communication, 27 November 2018). Within Public Health England, the Climate Change Group cooperates with the Extreme Events & Health Protection Group, which is based under the Environmental Hazards Department. Public Health England's goals and objectives in the field of climate change and health often originate in the

Climate Change Group and are subsequently passed on to higher levels within the organization before they arrive at the Health Department, which again coordinates with DEFRA before the proposal goes to the ministers (Oxford University, personal communication, 27 November 2018).

Overall, according to experts in the ministries, within the last eight years, the UK government has had lots of changes in the field of climate change, and an increasing number of officials is available to deliver policies and projects (DEFRA, personal communication, 28 November 2018). At the same time, however, many departments do not have the resources to do more than they are currently doing, although there is a desire to increase their adaptation efforts (DEFRA, personal communication, 28 November 2018).

Changes over time

As Figure 8.4 shows, it took until the release of the report "Health Effects of Climate Change in the UK 2012" in 2012 for the UK's health adaptation measures to really take off and, even then, the majority of reported measures were groundwork initiatives (Vardoulakis and Heaviside 2012). The largest increase in reported measures occurred in 2013 with the release of the National Adaptation Programme, in which the UK reached almost a perfect recognition score (6.5 out of 7 points), a comparatively high groundwork score (16.5 points), and the highest adaptation score in the country's history (58). The number of groundwork-level initiatives peaked in 2012 with a score of 25, rapidly decreased with the sixth NC to the UNFCCC and rose to 16.5 points again with the National Adaptation

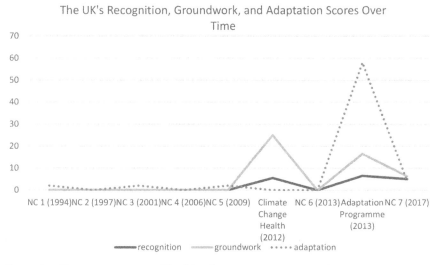

Figure 8.4 Time comparison of health adaptation measures in the UK

Programme from 2013. The high groundwork score in 2012 is mostly due to an exceptionally high number of recommendations (76 counted recommendations), whereas the National Adaptation Programme in 2013 does not include any recommendations on climate change and health but a large number of strategies and plans (25 counted strategies and plans).

Interestingly, the reporting of health adaptation initiatives in the NCs to the UNFCCC is constantly at a very low level. Only the NC7 from 2017 mentions more than one adaptation initiative and, compared with the initiatives mentioned in the other national documents, the reported measures in the NCs are extremely low.

Drivers and barriers

Climate change and health in the UK – risk assessment

According to the ND-GAIN index, the UK is among the least vulnerable countries to climate change in the world (ND-GAIN 2019b). Due to its comparatively low vulnerability score and its high readiness score, the UK represents the eight least vulnerable country on the ND-GAIN index (ND-GAIN 2019a). Health specific indicators on the ND-GAIN index demonstrate that challenges still exist, yet the high readiness of the country shows that it is generally better prepared to deal with risks stemming from climate change than other countries (see Table 8.1).

The health specific ND-GAIN data shows that the UK does not face any health risks due to a change of cereal yields, and only a very small fragment of the population will likely have to deal with deadly climate change related diseases (ND-GAIN Score of 0.030 compared with 0.303 for Sri Lanka or 0.333 for Sierra Leone). At the same time, the UK faces a high projected change in vector-borne diseases (0.658), a high flood hazard (0.735), and a relatively high impact of sea level rise (0.324).[2] As a consequence, it is clear that the UK has to adapt to climate change related health risks with a focus on preventing the spread of infectious diseases, preparing against health risks associated with extreme weather events, especially floods, and reacting

Table 8.1 ND-GAIN scores for climate change related health risks in the UK

Risk	ND-GAIN score
Projected change of cereal yields	0
Projected change of deaths from climate change induced diseases	0.030
Projected change in vector-borne diseases	0.658
Projected change of warm periods	0.114
Projected change of flood hazard	0.735
Projected change of sea level rise impacts	0.324

to the consequences of sea level rise. Other climate change related health risks, such as heat-related risks (0.114 ND-GAIN score in the category "projected change of warm periods"), should not be neglected. However, according to ND-GAIN, they play a less prominent role than other risks.

The analysis of data from the World Bank Climate Change Knowledge Portal leads to similar findings. Depending on the emission scenarios, the probability of an annual heatwave between 2040 and 2059 is 0.04 (RCP 4.5) or 0.06 (RCP 8.5). Although the probability rises the further the projection goes into the future, overall, the probability of a heatwave is in the UK still much lower than in other countries. The probability of a cold wave is projected to remain at −0.01 between 2040 and 2059 and then slightly decrease to −0.02 (RCP 4.5). Under a high emission scenario, the probability of a cold wave is to increase even faster, to −0.02 between 2040 and 2059, and then remain steady (RCP 8.5). This shows that, based on the projections, the UK will likely face slightly fewer cold-related health risks, such as death caused by hypothermia etc. The annual severe drought likelihood is not projected to largely increase in a medium-low emission scenario (RCP 4.5). Under a high emission scenario, the projected annual severe drought likelihood will increase to 0.09 by 2040–2058 and 0.10 by 2060–2079 (RCP 8.5), which is still a rather moderate increase compared with other countries.

Based on the results of the ND-GAIN index as well as the World Bank projections, it can be summarized that the major climate change related health risks for the UK are related to floods and storms, sea-level rise, and infectious diseases. Heat-related diseases may become an issue as well, but other countries will be significantly more affected. Information on secondary and tertiary climate change related health risks is rather scarce, which is why no qualified projections on allergies, mental health risks, migration, and conflicts can be made. Based on current research, it can be expected that especially mental health risks and allergies will become increasingly relevant.

Risks assessment by experts from the UK

According to top experts on climate change and health in the UK, the major climate change related health risks for the country are related to extreme weather events and extreme temperature, especially floods and heatwaves (LSHTM, personal communication, 25 November 2018; Leeds University, personal communication, 29 January 2019). This goes along with the findings of the UK Climate Change Risk Assessment 2017 Synthesis Report, which identifies an urgent need to act against floods and coastal change, health risks due to high temperatures, and, to a lesser extent, risks related to water and sanitation, food and agriculture, and new and emerging diseases (Committee on Climate Change, (CCC) 2016, 2). Moreover, risks related to food security and infectious diseases may become more serious as climate change will gain speed (LSHTM, personal communication, 25 November 2018). At the same time, an expert from Leeds University noted that the number and complexity of climate change related health risks in the UK is much larger (Leeds University, personal communication, 29 January 2019). Air quality related risks or those related to cardio-vascular and respiratory diseases will, for instance, become more prevalent in a changing climate and many risks have not yet been identified (Leeds University, personal communication, 29 January 2019).

Risk perception and the political agenda

For the last few years, BREXIT discussions have dominated the political agenda, making it difficult for other topics, such as climate change, to receive enough political attention (DEFRA, personal communication, 16 January 2019; Oxford University, personal communication, 27 November 2018). In the earlier 2000s, climate change was prominent on the agenda (DEFRA, personal communication, 16 January 2019). This was at least partly due to the massive heatwaves in the UK in 1996, 2003, and 2006, which were associated with climate change and led to heightened societal and political attention (DEFRA, personal communication, 28 November 2018; DEFRA, personal communication, 16 January 2019). Policies and institutions to combat climate change and its effects really developed after the 2003 heatwave, which many experts described as a "wake up call in all Europe" (Leeds University, personal communication, 29 January 2019). The UK, for instance, started to develop its national heatwave plan as a reaction to the extreme temperatures and, overall, climate change was high on the agenda (Leeds University, personal communication, 29 January 2019). The awareness and actions culminated in the Climate Change Act from 2008, which led to the establishment of the British Framework on Climate Change, the creation of various institutions that work on the different aspects of climate change and health, and overall widespread recognition that both adaptation and mitigation are needed in the UK to respond to climate change related risks (DEFRA, personal communication, 28 November 2018).

After 2008, climate change gradually lost attention during austerity and became more contested (DEFRA, personal communication, 16 January 2019). Nevertheless, unlike in other countries, there is not much climate change denial and a cross-party consensus exists on the fact that climate action is necessary (DEFRA, personal communication, 16 January 2019). One official from DEFRA stated: "Climate skepticism around science has been put to bed" (DEFRA, personal communication, 16 January 2019). However, disagreements both in the media and politics persist regarding which specific measures should be taken (DEFRA, personal communication, 16 January 2019).

Climate change and health is still, at least from time to time, on the government's agenda, but the focus often lies on a small number of health risks rather than the complex relationship between climate change and health

(Leeds University, personal communication, 29 January 2019). Much of the debate focuses on heat-related health risks, while more diffuse risks, such as those related to air quality or cardio-vascular and respiratory diseases, are less discussed, although they will become more prevalent as climate change progresses (Leeds University, personal communication, 29 January 2019). Governmental experts confirmed that the major climate change related health risks that the government is concerned about are related to severe weather events, such as flooding, and extreme temperatures (DEFRA, personal communication, 28 November 2018). Especially the heatwaves have led to more attention for the health effects of climate change and, therefore, the prompted government to foster research on the topic to reduce uncertainty and build adaptive capacity (DEFRA, personal communication, 28 November 2018).

In 2018 and the beginning of 2019, climate change has received more awareness again, as extreme weather events, especially floods, hit the UK and pushed climate change more to the agenda (DEFRA, personal communication, 16 January 2019). During the 2018 heatwave, a new scientific report on heat effects was published and public awareness rose again. However, after some time, other topics pushed the topic away from the agenda (LSHTM, personal communication, 25 November 2018). Nonetheless, compared with around ten years ago, adaptation has gained much more relevance and attention, and institutions such as the Adaptation Sub-Committee have contributed to raising awareness for adaptation, engaging with civil society, mainstreaming adaptation across different policy areas, and monitoring the government's progress in the field (LSHTM, personal communication, 25 November 2018). Recently, mitigation and adaptation have become more separated in public debate, but, unlike in other countries, climate change adaptation does not have a negative image in the UK (LSHTM, personal communication, 25 November 2018).

Dealing with complex decisions under uncertainty

The nexus between climate change and health is very complex and still some uncertainty exists in terms of the concrete climate change related health risks that are and will become relevant in the UK, as well as the scale and intensity of the risks (DEFRA, personal communication, 16 January 2019). Therefore, universities and research institutes play an important role for the government's health adaptation efforts. According to a governmental official, especially senior academics and larger research centers have a lot of influence on climate change adaptation in the UK (DEFRA, personal communication, 16 January 2019). At the same time, some interviewed scholars claimed that more interdisciplinary research was necessary to develop an even better understanding of the issue (Leeds University, personal communication, 29 January 2019).

It is worth noting that the UK's level of research on the topic is far higher than that of other countries and, after the heatwaves in 1996 and 2003, the government has financed many research projects on the impacts of climate change, including the health sector (DEFRA, personal communication, 28 November 2018). Academia plays an especially important role for governmental agencies, such as the Health Protection Agency (LSHTM, personal communication, 25 November 2018; Oxford University, personal communication, 27 November 2018). Consequently, the UK is much better prepared to deal with the uncertainty connected to climate change and can make more informed decisions, which can at least partially explain the UK's high ranking on the CHAIn. As experts from the LSHTM stated, for policymakers it is important to have a strong evidence base, and in the UK research findings now show the impact of climate change on people's health more clearly than ever (LSHTM, personal communication, 25 November 2018). Not only is this important for decision makers to identify current priorities, but it also helps them to measure a "return of investment" (LSHTM, personal communication, 25 November 2018). Especially the Climate Change Act from 2008 was a starting point for capacity building both in the public sector and with regards to funding research projects in the field (Oxford University, personal communication, 27 November 2018). Moreover, research institutes in the UK benefited from EU funding on climate change and health (LSHTM, personal communication, 25 November 2018).

Governmental institutions in the field of climate change have undergone many changes in the past eight years and more officials are now working on climate change related topics than before (DEFRA, personal communication, 28 November 2018). The link between academia and politics has improved over time and more research councils have been established to provide expertise on the specific impact of climate change in the UK (DEFRA, personal communication, 28 November 2018). Additionally, an increasing number of interdisciplinary and cross-sectoral projects has been conducted (DEFRA, personal communication, 28 November 2018).

The social and cultural dimension

In the UK, the public is aware of and well-informed about climate change in general, but less so about the various health effects of climate change (Leeds University, personal communication, 29 January 2019). Especially when it comes to new emerging diseases as a consequence of climate change, the public understanding is rather limited (Leeds University, personal communication, 29 January 2019). Although, according to experts, health is not the first thing that comes to people's minds when they think about climate change, numerous civil society organizations have gathered support for climate action and raised awareness for the health effects of climate change in the UK (LSHTM, personal communication, 25 November 2018). Their influence largely varies over time, but there are Non-Governmental

Organizations (NGOs) that put pressure on the government by organizing demonstrations and raising awareness through media campaigns (DEFRA, personal communication, 28 November 2018; Leeds University, personal communication, 29 January 2019). One example is the Extinction Rebellion, which protests to raise awareness for climate change by blocking bridges and claiming that politicians are killing humanity with their inaction (BBC 2019). Friends of the Earth, the Global Action Plan, Client Earth, and others also strive to raise awareness and push for stronger governmental climate action (DEFRA, personal communication, 28 November 2018; Oxford University, personal communication, 27 November 2018). Moreover, specific health-related NGOs and advocacy groups exist. The UK Health Alliance on Climate Change advocates for more health specific mitigation and adaptation measures as well as an even stronger research basis (HACC 2019) The Lancet Countdown on Climate Health and Climate Change unites researchers, representatives of international organizations, and activists, and publishes annual reports on the effects of climate change on health across the globe (LSHTM, personal communication, 25 November 2018).

The influence of civil society organizations peaked with the passing of the Climate Change Act in 2008 and, since then, it rises and declines in relation to other developments, such as major climate change negotiations and agreements or extreme weather events (DEFRA, personal communication, 16 January 2019). Environmental NGOs mostly receive media attention when international climate change related events happen or specific projects, such as the expansion of airports, are discussed (DEFRA, personal communication, 16 January 2019). Now with BREXIT dominating the discussions, more and more NGOs lobby for making the exit as green as possible and see it as an opportunity to redirect public money towards protecting the environment (DEFRA, personal communication, 16 January 2019).

The economic dimension

Although the UK is economically well situated and, compared with other countries, has enough capacity to invest in long-term challenges such as climate change, general economic trends, such as the European economic crisis that began to unfold in 2008 and the national austerity policy, had an effect on health adaptation to climate change in the UK (DEFRA, personal communication, 16 January 2019; Leeds University, personal communication, 25 January 2019). During austerity, other topics dominated the agenda and climate change received less attention (DEFRA, personal communication, 16 January 2019). Moreover, due to cutbacks as part of the austerity policies, the health care sector has received less funding in recent years, although with regards to the UK's demographic development, even more investments are necessary (Leeds University, personal communication, 25 January 2019). According to an expert on social environmental policies, this has led to a social dilemma where the willingness to pay for health care is not there and the UK's health system is under a lot of pressure (Leeds University, personal communication, 25 January 2019). Due to these immediate and mediate challenges, climate change is not on the forefront of the agenda of the health sector (Leeds University, personal communication, 25 January 2019).

At the same time, when the adaptation framework was institutionalized in the early 2000s, local governments were given major responsibilities and funding to work on adaptation, especially with regards to flooding and health risks (Leeds University, personal communication, 25 January 2019). During austerity, however, the government abolished the old adaptation system and the local government level lost a large part of its funding (Leeds University, personal communication, 25 January 2019). Since other public services were more pressing, the local entities had to cut their efforts on climate change and, as a consequence, policy-dismantling took place (Leeds University, personal communication, 25 January 2019). This was less an issue for mitigation since it had been enshrined in law, but adaptation really suffered from the austerity policies (Leeds University, personal communication, 25 January 2019). Still, many mechanisms and institutions, such as he Adaptation Sub-Committee are still in place and work effectively (Leeds University, personal communication, 25 January 2019).

Both in politics and the general public, the economy is often at the top of the agenda, whereas the environment is more at the bottom (DEFRA, personal communication, 16 January 2019). People often care more about specific risks, such as those related to floods or heat (DEFRA, personal communication, 16 January 2019). If asked about these risks, environmental issues often jump right to the top of their priorities (DEFRA, personal communication, 16 January 2019). Moreover, an increasing number of people makes the connection between climate change and health, which often leads to more awareness for climate change in general (DEFRA, personal communication, 16 January 2019). On the ministerial level, however, the link between climate change and health is less prominently discussed (DEFRA, personal communication, 16 January 2019).

The international dimension

The majority of interviewed experts has claimed that the direct influence of international organizations, such as the WHO, is rather limited in the UK since these organizations are more active in developing countries (LSHTM, personal communication, 25 November 2018). Although the WHO's office in Bonn has a focus on climate change and health and provides technical support to European countries, it does not get much involved in developing policies with and for European countries since it focuses more on technical details (LSHTM, personal communication, 25 November 2018). In developing countries, on the other hand, the WHO, other international organizations, or even single countries, have a strong influence on health adaptation policies (LSHTM, personal communication, 25 November 2018). Germany, for instance, has funded several National Adaptation Plans for developing countries (LSHTM, personal communication, 25 November 2018). This observation goes along with the finding from the longitudinal assessment of the NCs to the UNFCCC, which shows that the UK does not report as extensively on their health adaptation policies in their NCs than in their national-level documents, such as the National Adaptation Programme.

Despite the comparatively low direct influence, international organizations definitely have a strong indirect influence on health adaptation in the UK. International events, such as Conferences of the Parties to the UNFCCC (COPs) or G7 and G20 meetings, can bring the topic to the agenda and put pressure on governments to act against climate change (DEFRA, personal communication, 16 January 2019; Oxford University, personal communication, 27 November 2018). The UK often plays an important role in shaping international climate change agreements and increasing global ambition, as was the case especially with the Paris Agreement in 2015 (DEFRA, personal communication, 16 January 2019). The UK tends to strive for international reputation at international climate change conferences and is interested in coming across as a very engaged climate leader (DEFRA, personal communication, 16 January 2019). Overall, the UN and its various institutions have strong legitimacy in the UK, and especially the WHO is frequently mentioned in governmental decisions (DEFRA, personal communication, 16 January 2019). The WHO's benchmarks and reports are usually considered by British policymakers, which leads to what one of the interviewed experts called "soft power" through their legitimacy and through shaping the discourse (DEFRA, personal communication, 16 January 2019). Moreover, the WHO often sets concrete standards that are picked up by decision makers (DEFRA, personal communication, 16 January 2019). As a consequence, the WHO is one of the most respected and talked about UN organizations in the UK (DEFRA, personal communication, 16 January 2019). The climate change debate is slightly more contested, which is why, from the perspective of a DEFRA official, the UNFCCC is less influential than the WHO (DEFRA, personal communication, 16 January 2019). UN Environment is even less acknowledged than the UNFCCC since many policymakers prefer to set their own environmental standards rather than adhering to international ones (DEFRA, personal communication, 16 January 2019). The IPCC's publications, on the other hand, are broadly received, and especially the recent special report "Global Warming of 1.5 °C" has helped to set climate change on the agenda and put pressure on policymakers in the UK to increase their adaptation and mitigation efforts (DEFRA, personal communication, 28 November 2018).

The EU also plays an important role for the UK's adaptation policies, as it sets standards and requires reporting in many areas (LSHTM, personal communication, 25 November 2018). The EU Climate Adapt program and the EU's questionnaire on climate change adaptation are of importance for the UK's adaptation policies since they require them to summarize their efforts and report them to the EU (Oxford University, personal communication, 27 November 2018). Moreover, the EU's funding for research projects in the UK contributed to the development of a strong research basis for the government's decisions on the topic (Oxford University, personal communication, 27 November 2018).

Furthermore, a number of experts stated that cooperation between the UK and other states is rather limited and mostly occurs when governmental delegations visit the UK (DEFRA, personal communication, 16 January 2019; Oxford University, personal communication, 27 November 2018). Nevertheless, in some cases, governmental agencies, such as Public Health England, develop projects with foreign partners, as was the case for a project in Cyprus (Oxford University, personal communication, 27 November 2018). A few years ago, Public Health England conducted a project in Cyprus where they developed a heatwave preparedness plan for this Mediterranean country and launched first research projects on climate change and health (Oxford University, personal communication, 27 November 2018). Clearly, this form of cooperation is more beneficial to the partner country, in this case Cyprus, but it shows that policy diffusion from the UK to other countries takes place. Moreover, the high rank of Cyprus may at least partially be explained through the support they received.

Overall, international organizations are definitely an important player in the UK since they work closely with academia, shape the discourse, raise awareness, and set standards (DEFRA, personal communication, 16 January 2019). Unlike in developing countries, however, there is usually no concrete support for project development and implementation within the UK since the UK has been a leader on climate change and has a lot of own research capacity (DEFRA, personal communication, 16 January 2019).

Summary

The UK's strong performance on the CHAIn can be traced back to the lessons learned from extreme weather events in the early 2000s, especially the heatwave from 2003. Although the UK was and is not significantly more affected by climate change than other countries, the topic was high on the political agenda between 2003 and 2007 and civil society's pressure on the government to act against climate change and its consequences significantly rose. The awareness for climate change culminated in the adoption of the Climate Change Act in 2008, which led to the establishment of stronger institutions and higher capacities in the adaptation sector. Especially public agencies, such as Public Health England, established working groups on climate change and health. After the economic crisis in 2008 and the resulting austerity policy, climate change received much less attention and since then the topic mostly gets on the agenda when important international conferences happen and groundbreaking agreements, such as the Paris Agreement (2015), are adopted. Moreover, funding for adaptation has been cut, especially on the local level, and additional pressure on the health system has led to a change of priorities and less attention for climate change related health risks. Since the UK's vote to leave the EU on 23 June 2016, the political agenda has been dominated by Brexit discussions and climate change has received less awareness. Civil society organizations used to be rather strong in the early 2000s and now rise every now and then, as the example of the Extinction Rebellion shows. Their presence often correlates with international developments or climate change related events within the country, such as discussions on expansions of airport runways.

Nevertheless, the institutions that developed after the adoption of the Climate Change Act as well as the strong research basis in the UK, with the world's best universities and research institutes, help the country to reduce some of the uncertainty and complexity that the nexus between climate change and health involves. As a consequence, much more is known about the specific climate change related health risks in the UK than in other countries, and it is easier for decision makers to justify their actions since they can more easily than elsewhere show a return of investment. The UK's reaction to the heatwave in 2003 and other events differed from that of other European countries, especially since it led to a new wave of institutionalization of climate action, from which the UK's health adaptation work still benefits. This shows that the perception and policy learning from extreme weather events really differs from country to country and it is not just the exposure or vulnerability to such events, but it is key to have a political environment that enables policy learning and the necessary knowledge and expertise to develop solutions.

Notes

1 All of these documents were identified as part of the CHAIn analysis.
2 The Projected change of cereal yields describes the "projected impact of climate change on the actual yields of rice, wheat and maize. Projected change is the percent decrease of the cereal yields from the baseline projection (1980–2009) to a future projection (2040–2069) using RCP 4.5 emission scenario" (ND-GAIN 2019b). The projected change of flood hazard implies "Projected impact of the climate change on flood. The flood hazard is measured by the monthly maximum precipitation in 5 consecutive days. The projected change is the percent increase of the flood hazard from the baseline projection (1960–1990) to the future projection (2040–2070), using RCP 4.5 emission scenario" (ND-GAIN 2019b).

References

BBC. 2019. "Extinction Rebellion protests: what happened?" *BBC Online*, 25 April. https://www.bbc.co.uk/news/uk-england-48051776.
CCC. 2016. *UK Climate Change Risk Assessment 2017*. London: Committee on Climate Change.

The UK and Northern Ireland 149

HACC. 2019. "About." Accessed 25 May 2019. http://www.ukhealthalliance.org/about/.

ND-GAIN. 2019a. "The ND-GAIN Matrix." Accessed 23 May 2019. https://gain-new.crc.nd.edu/matrix.

ND-GAIN. 2019b. "United Kingdom." Accessed 23 May 2019. https://gain.nd.edu/our-work/country-index/.

Vardoulakis, Sotiris, and Clare Heaviside. 2012. *Health Effects of Climate Change in the UK 2012*. London: Health Protection Agency.

9 The Republic of Ireland

Index test

The expert interviews have shown that Ireland's generally low CHAIn score can be confirmed since the most important documents on national-level health adaptation to climate change in the country had been identified and analyzed in the LNA. They include the NCs and the National Climate Change Adaptation Framework, the Climate Action and Low Carbon Development Act from 2015, and the Environmental Protection Agency's Climate Change Vulnerability Assessment.[1] According to an influential member of Ireland's adaptation community, in recent years the country has focused on preparing the policies and actions that will be launched in the near future (MaREI Centre, personal communication, 23 November 2018). It can therefore be expected that Ireland's CHAIn score will increase soon. At the same time, the discussions with experts have indicated that Ireland's national-level documents do not include all health adaptation measures that the country has developed over the years, as some state-funded research projects or capacity building measures do in fact exist but are not mentioned in the reports. This shows that the index score does not always represent what is happening on the ground, but rather indicates how governments interpret health adaptation, what they associate with it, and what they deem necessary to mention. Consequently, despite some imperfections, the index allows many conclusions about the importance of health adaptation to climate change in Ireland and the progress in the field. Moreover, the interviews have confirmed that many experts agree that, compared with other European countries, Ireland is in fact late in adapting to climate change related health risks, which is reflected in the low index scores (Department of Health, personal communication, 22 November 2018; Technical University of Dublin, personal communication, 24 November 2018).

Governance

This section describes Ireland's health adaptation measures to climate change in international comparison, the different measures the government has taken, as well as the institutional framework and the key players in the Irish adaptation community.

International comparison

With an overall CHAIn score of 9, a recognition score of 6.5, a groundwork score of 0.5, and an adaptation score of 2, Ireland ranks at place 123 on the CHAIn. It is followed by Paraguay (CHAIn score of 9, but no reported adaptation actions), Zambia (CHAIn score of 8.5), Afghanistan (8.25), the Bahamas (8.25), and the Gambia (8.25). Whilst some other European countries score even lower on the CHAIn, such as Latvia (rank 129 with a CHAIn score of 8.25), Slovenia (rank 130 with a CHAIn score of 8.25), or Luxembourg (rank 165 with a CHAIn score of 4.75), Ireland clearly ranks at the lower midfield of the CHAIn. Many developing countries, such as Kiribati (rank 120 with a CHAIn score of 9.25), Rwanda (rank 34 with a CHAIn score of 25.25) or the Solomon Islands (rank 7 with a CHAIn score of 49.25), rank significantly higher than Ireland.

Health adaptation portfolio

Although Ireland has a very high recognition score (6.5 out of 7), which indicates that the country acknowledges all climate change related health risks that are prominently discussed in academia, except for those related to conflicts, its groundwork-level and adaptation-level initiatives are at a very low level. As Figure 9.1 shows, in its official national documents Ireland only reported one general groundwork initiative and no groundwork measures on any specific health risks.

Similarly, only one adaptation initiative that addresses general climate change related health risks was identified (Figure 9.2).

The two general measures include one groundwork initiative in the category "strategies & plans" and one adaptation initiative in the category

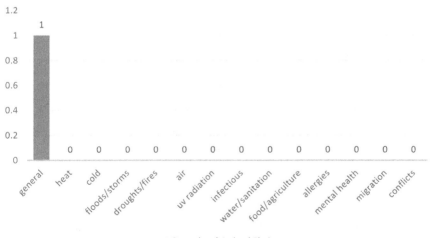

Figure 9.1 Groundwork-level initiatives, Ireland

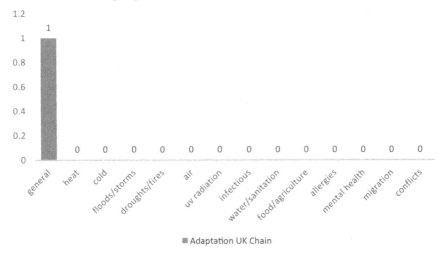

Figure 9.2 Adaptation-level initiatives, Ireland

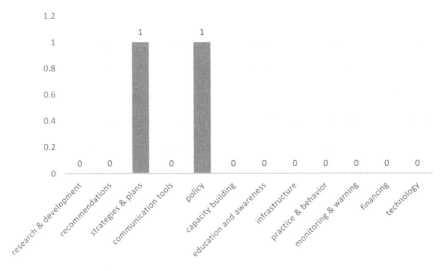

Figure 9.3 Adaptation types, Ireland

"policy." No research projects, recommendations, or communication tools were mentioned in the national-level documents (see Figure 9.3).[2]

Institutional framework and key actors

Ireland's institutional framework in the field of climate change adaptation really developed in the 2010s, especially when the Non-Statutory National Climate Change Adaptation Framework (NCCAF) was established in 2012, the EU Adaptation Strategy was released in 2013, and the Climate Action and

Low Carbon Development Act was adopted in 2015. The Climate Action and Low Carbon Development Act has started a number of adaptation strategies and actions (Department of Health, personal communication, 22 November 2018). It officially requested the Department of Communications, Climate Action and Environment Ireland (DCCAE) to develop a climate change adaptation plan (Department of Health, personal communication, 22 November 2018). Until the end of September 2019, all sector specific adaptation plans for 2019–2024 will be published (Department of Health, personal communication, 22 November 2018). The plans will be updated at least every five years and the respective ministers need to sign off the plans that have been developed by their ministerial officials who deal with climate change (Department of Health, personal communication, 22 November 2018).

The role of the different ministries in terms of adaptation to climate change in Ireland is quite clear (MaREI Centre, personal communication, 23 November 2018). The DCCAE is in charge of Ireland's overall adaptation initiatives to climate change and the Department of Health is responsible for health-specific adaptation measures (DCCAE, personal communication, 22 November 2018; Department of Health, personal communication, 22 November 2018). DCCAE has issued guidelines for the different governmental sectors to develop sector specific plans, which are similar to what the European Commission has introduced under the EU Adaptation Strategy (DCCAE, personal communication, 22 November 2018). DCCAE coordinates and integrates the sector specific plans into a comprehensive strategy and national plan (DCCAE, personal communication, 22 November 2018). The new national adaptation plan is scheduled to be published in September 2019 (DCCAE, personal communication, 22 November 2018).

In practice, the Department of Health cooperates very closely with the Health Service Executives (HSE) of Ireland (Department of Health, personal communication, 22 November 2018). Moreover, several research institutes, universities, and governmental agencies, such as the Institute of Public Health and the Environmental Protection Agency, play an important role for the ministry's work in the field (Department of Health, personal communication, 22 November 2018).

Over the years, numerous councils and assemblies have been established to deal with adaptation to climate change in Ireland, such as the Climate Change Advisory Council, which was established under the Climate Action and Low Carbon Development Act (Kelly, personal communication, 23 November 2018). Ireland's Environmental Protection Agency supports the work of the government and the Climate Change Advisory Council through various research projects (Kelly, personal communication, 23 November 2018). In addition to cooperating with academia, governmental agencies and sub-organizations conduct their own research on climate change and report to ministerial entities (DCCAE, personal communication, 22 November 2018). Their work on climate change and health is, however, rather limited (DCCAE, personal communication, 22 November 2018).

Moreover, the National Climate Dialogue and the Citizens' Assembly have been established to strengthen Ireland's climate change strategies and to connect civil society organizations, individual citizens, and policy-makers (Kelly, personal communication, 23 November 2018). Last but not least, the National Adaptation Committee oversees the national adaptation progresses and integrates key stakeholders from politics, academia, the private sector, and civil society (Department of Health, personal communication, 22 November 2018). It evaluates Ireland's progress on climate change adaptation and makes proposals on how the country's adaptation work can be improved (Department of Health, personal communication, 22 November 2018).

Development over time

According to the official national-level documents, climate change and health was not really addressed in Ireland's climate change documents until the publication of the National Adaptation Plan in 2012, which led to a strong increase in the country's recognition score (6 out of 7 points). Subsequent documents show a significantly lower recognition score (only 3 out of 7 points in the Vulnerability Assessment from 2013 and no explicitly recognized risks in the subsequent NCs or the Climate Action and Low Carbon Development Act).

The only groundwork-level measure mentioned was reported in the most recent NC from 2018 and the only adaptation-level measure was found in the Climate Action and Low Carbon Development Act from 2015. Overall, Ireland's health adaptation work is constantly at a rather low level and under-reported in the NCs to the UNFCCC.

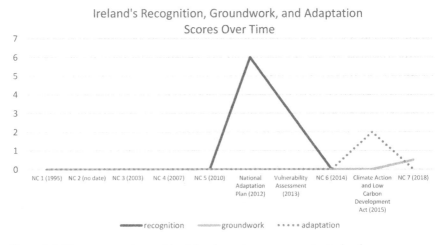

Figure 9.4 Time comparison of health adaptation measures in Ireland

Drivers and barriers

Climate change and health in Ireland – risk assessment

Ireland has a comparatively low vulnerability to climate change. According to the ND-GAIN index, Ireland is the twenty-eighth least vulnerable country to climate change and the twenty-second most ready country to deal with the consequences of climate change (ND-GAIN 2019). When looking at health-specific indicators, it becomes clear that Ireland will face a number of serious challenges as climate change progresses, but, compared with other countries, it is less vulnerable.

The health-specific index scores show that Ireland will only face a slight increase in deaths from climate change induced diseases (ND-GAIN Score of 0.030), but a relatively high change in vector-borne diseases (0.658) and a high projected change of flood hazard (0.737). The projected change of warm periods (0.113) and sea level rise impacts (0.104) are much lower than in other countries, but still should not be neglected. As a consequence, the ND-GAIN data suggests that the major climate change related health risks for Ireland are related to vector-borne diseases and flood hazards.

The data from the World Bank Climate Change Knowledge Portal goes along with these findings. For Ireland, the probability of an annual heatwave between 2040 and 2059 ranges between 0.04 (RCP 4.5 – medium-low emission scenario) and 0.07 (RCP 8.5 – high emission scenario). This suggests that, whilst the probability of heatwaves will rise with climate change, the increase is less dramatic than in other countries. The annual probability of a cold wave is projected to remain stable at −0.01 between 2040 and 2059 under a medium-low emission scenario (RCP 4.5), whereas it is projected to decrease to −0.02 in a high-emission scenario (RCP 8.5). This shows that, based on the World Bank's climate projections, Ireland can expect a slight reduction in cold-related health risks as climate change progresses. The annual severe drought likelihood is projected to slightly decrease until 2040–2059 to 0.04 and then slightly increase to 0.08 for 2060–2079 under a medium-low emission scenario (RCP 4.5). Under a high-emission scenario

Table 9.1 ND-GAIN scores for climate change related health risks in Ireland

Risk	ND-GAIN score
Projected change of cereal yields	No data
Projected change of deaths from climate change induced diseases	0.030
Projected change in vector-borne diseases	0.658
Projected change of warm periods	0.113
Projected change of flood hazard	0.737
Projected change of sea level rise impacts	0.104

(RCP 8.5), the annual severe drought likelihood is projected to slightly increase to 0.08 until 2040–2059, then decrease to 0.07 for 2060–2079 and increase to 0.11 for 2080–2099. The World Bank's data suggests that Ireland will probably not face severe health risks due to droughts in the near future.

To summarize, both the data from ND-GAIN and the World Bank suggest that the major climate change related health risks that Ireland will face are related to extreme weather events, such as floods. Sea-level rise and vector-borne diseases also constitute significant risks. Although Ireland is not completely safe from heat-related diseases or other climate change related health risks, it is less vulnerable in this field than other countries.

Risks assessment by experts from Ireland

According to Irish experts on the topic, the major climate change related health risks for the country are connected to severe weather events, such as floods and storms and sea-level rise (Cullen 2007). In 2017, with hurricane Ophelia, Ireland experienced its worst storm in 50 years (MaREI Centre, personal communication, 23 November 2018; Sweeney, personal communication, 23 November 2018). Ophelia caused severe damage to the country's infrastructure and killed three people (Sweeney, personal communication, 23 November 2018). In addition to the direct effects of the storms on people's health, Ireland has experienced a change in risk-taking behavior: in the middle of storm Ophelia in 2017, a man jumped into the ocean and thereby risked not only his life but also the lives of those of the emergency response team (Dinham 2017).

Moreover, heat-related health risks pose severe challenges to the Irish population (Kelly, personal communication, 23 November 2018; Technical University Dublin, personal communication, 24 November 2018). In 2018, Ireland experienced one of its hottest and driest summers in the country's history (Sweeney, personal communication, 23 November 2018). Although other countries are exposed to much higher temperatures than Ireland, the Irish population is not used to heat and therefore is less prepared to deal with heatwaves (Sweeney, personal communication, 23 November 2018; Technical University Dublin, personal communication, 24 November 2018). According to an expert, temperatures over 25 °C are considered as heat in Ireland and temperatures over 30 °C are considered as a heatwave (Technical University Dublin, personal communication, 24 November 2018). Moreover, in Ireland, many houses are not properly insulated and prepared for extreme temperatures (Sweeney, personal communication, 23 November 2018). With regards to extreme temperature, Ireland will probably witness slight reductions in mortality rates due to reduced winter mortality, although summer mortality may rise (Technical University Dublin, personal communication, 24 November 2018).

Climate change will make it more and more difficult for Ireland to ensure high water quality and supply, especially when heatwaves and droughts occur more frequently (Department of Housing 2019). With warming winters, more food- and waterborne diseases can spread since pathogens do not get killed with low temperatures anymore (Kelly, personal communication, 23 November 2018; Sweeney, personal communication, 23 November 2018). Additionally, increased precipitation can help organisms to be transported over longer distances and thus get consumed by more people (Kelly, personal communication, 23 November 2018). Food contamination, salmonella, and waterborne diseases, such as Verocytotoxigenic Escherichia coli (VTEC), already occur more often in Ireland, especially since 11 percent of the population relies on private wells that are less frequently checked and up to 30 percent of private wells are prone to E. Coli, cryptosporidium, and other pathogens (Kelly, personal communication, 23 November 2018; MaREI Centre, personal communication, 23 November 2018; Sweeney, personal communication, 23 November 2018). Ireland has by far the highest rates of confirmed Verotoxigenic E. coli cases in Europe, which is at least partly due to the water governance system in Ireland, which allows less quality controls for privately owned wells (Kelly, personal communication, 23 November 2018).[3]

Additionally, infectious diseases, such as malaria, blue tongue, and other diseases may spread to new regions, and new species may invade Ireland (Technical University Dublin, personal communication, 24 November 2018). At the moment, however, infectious diseases are less a concern than risks related to extreme weather events or extreme temperature (Technical University Dublin, personal communication, 24 November 2018). Additionally, mental health risks related to extreme weather events will likely increase as more and more people will lose their properties due to floods, storms, and other severe weather events (MaREI Centre, personal communication, 23 November 2018).

Risk perception and the political agenda

Climate change is a frequently discussed topic in Ireland, especially after extreme weather events occur (MaREI Centre, personal communication, 23 November 2018). The population generally recognizes that anthropogenic climate change is happening and only a few politicians deny climate change (Institute of Public Health, personal communication, 22 November 2018). Mitigation has been on the agenda for many years and now with more extreme weather events, the need for adaptation is discussed more often as well (MaREI Centre, personal communication, 23 November 2018). The term adaptation does not have a negative image, but the focus has been more on mitigation in the past (MaREI Centre, personal communication, 23 November 2018). Nowadays, adaptation is often easier to communicate and implement in Ireland because there is less resistance than with mitigation (Department of Health, personal communication, 22 November 2018). Compared with other topics, such as the impact of climate change on coastal areas, the nexus between climate change and health is, however, not that prominent on

Ireland's adaptation agenda (Institute of Public Health, personal communication, 22 November 2018; MaREI Centre, personal communication, 23 November 2018). At the same time, while health adaptation is often a continuation of existing work in the field, the link to climate change is not made (Department of Health, personal communication, 22 November 2018).

Ireland's geographical location and weather conditions lead to lots of discussions about the weather and the climate and, especially after extreme weather events happen, policy learning takes place (Department of Health, personal communication, 22 November 2018; Technical University of Dublin, personal communication, 24 November 2018). After such events, societal and media pressure put the topic high on the political agenda (Department of Health, personal communication, 22 November 2018; Technical University of Dublin, personal communication, 24 November 2018). In the past, the population often believed that Ireland was less affected by climate change than other countries, but now the public is starting to realize the scale and intensity of the impact of climate change in Ireland (Institute of Public Health, personal communication, 22 November 2018; Sweeney, personal communication, 23 November 2018). However, after a while, the topic often loses momentum and the agenda is dominated by other issues again (Technical University of Dublin, personal communication, 24 November 2018).

Additionally, more and more people are becoming fatigued by the topic of climate change, as there was a lot of media coverage in previous years and people have heard a lot about it (Institute of Public Health, personal communication, 22 November 2018). Importantly, when the economic crisis broke out in 2008, climate change was pushed to the back of the agenda and, instead of large-scale governmental actions, the focus of the debate was more on what individuals have to do, which may have led to even more fatigue and disappointment in the population (Institute of Public Health, personal communication, 22 November 2018).

Within the health discourse, numerous experts stated that the Irish health system has been in crisis for decades, especially in terms of the accident and emergency system (Technical University of Dublin, personal communication, 24 November 2018). Especially on weekends it can be very difficult for patients to get appointments with doctors and in hospitals people have to wait for days to get a bed (Technical University of Dublin, personal communication, 24 November 2018). Due to the austerity during the economic downturn and reduced pay for doctors and nurses, many health professionals have left the country (Kelly, personal communication, 23 November 2018). As a consequence, the government's key priority is to solve the health service crisis, and more long-term issues such as climate change receive less attention (Technical University of Dublin, personal communication, 24 November 2018). Although the country is now picking up in developing more comprehensive adaptation strategies, including the statutory sectoral plans in 2019, Ireland is still quite late in its groundwork-level adaptation work to climate change related health risks, which is at least partly due to lacking governmental support and capacity to develop health sectoral plans (Kelly, personal communication, 23 November 2018).

Dealing with complex decisions under uncertainty

Whilst the Department of Health and the Irish research community have started to examine the relationship between climate change and health, numerous knowledge gaps exist, especially when it comes to the concrete health effects of climate change in Ireland (Department of Health, personal communication, 22 November 2018; MaREI Centre, personal communication, 23 November 2018). Since health adaptation to climate change is a very complex challenge with many unknowns, the Irish government first has to develop their groundwork initiatives to get a grasp of the topic before adaptation actions can be implemented. Currently, the department's adaptation policies to climate change related health risks still depend on the expertise of research institutes, universities, and governmental agencies, such as the Institute of Public Health, the Health Service Executives (HSE), and the Environmental Protection Agency (Department of Health, personal communication, 22 November 2018; Institute of Public Health, personal communication, 22 November 2018). The departments of the institutes are, however, often divided according to the topics that were on the agenda ten or twenty years ago and are not appropriately updated to meet the needs of new and broader global challenges, such as climate change (Institute of Public Health, personal communication, 22 November 2018).

According to government officials, Ireland needs more cross-sectoral and interdisciplinary cooperation and more concrete information on which health risks the population will be exposed to as a consequence of climate change (DCCAE, personal communication, 22 November 2018; Department of Health, personal communication, 22 November 2018). Dr. Ina Kelly from HSE Midlands summarized the role of climate change adaptation and its needs in Ireland as follows:

> Climate Change adaptation allows us to fix a lot of problems that are already actual problems at the moment. Climate change adaptation allows us to shine a light on them and to show other sectors that they can work on them. It offers an opportunity for inter-sectoral analysis and engagement and addressing issues across sectors that siloed arrangements, that we usually have, don't allow. Climate change allows us to work better together.
> (Kelly, personal communication, 23 November 2018)

The Department of Health's own capacity to deal with the topic is very limited – only one part-time position is currently dedicated to climate change and health. Therefore, the cooperation with HSE and other health institutions is of outmost importance for the ministry (Department of Health, personal communication, 22 November 2018). Universities and research institutes complement the government's own research on climate change, and some flagship projects, such as the Irish Climate Analysis and Research UnitS (ICARUS) at Maynooth University, help the government to learn about risks that they cannot focus on in their operational work (DCCAE, personal communication, 22 November 2018; Kelly, personal communication, 23 November 2018). The government is, however, planning on expanding the ministry's capacities in the field of climate change and health (Department of Health, personal communication, 22 November 2018).

The Irish government has funded a number of research projects at universities focused on climate change, but compared with other topics, the capacity in the field is limited (Department of Health, personal communication, 22 November 2018). Moreover, cooperation between different universities, research institutes, and governmental institutes is often on a voluntary basis and not very widespread (MaREI Centre, personal communication, 23 November 2018). Recently established regional offices on climate change partner up with academia to develop research projects that address their needs (MaREI Centre, personal communication, 23 November 2018). The Environmental Protection Agency's (EPA) research program has guided academic research in terms of groundwork adaptation and preparing action in the future (MaREI Centre, personal communication, 23 November 2018). DCCAE has developed instruction manuals on how strategies and plans in all adaptation fields, including health, should be developed for the short-, medium-, and long-term future and, in 2019 Ireland's health adaptation framework will develop further (DCCAE, personal communication, 22 November 2018). Adaptation in other sectors, however, is much more straightforward and clear, which is one of the reasons why Ireland's initiatives in areas like flood control are ahead of health adaptation (DCCAE, personal communication, 22 November 2018).

Many experts claimed that the key challenge for effective mitigation and adaptation the Irish government has to overcome is to stop working in silos and to foster more cross-sectoral cooperation (Sweeney, personal communication, 23 November 2018). Often the different departments tend to defend their area of work, which prevents real collaboration across departments (Sweeney, personal communication, 23 November 2018). Another barrier to effective adaptation is the electoral system, which is based on multi-seat constituencies where up to five members for each constituency are rivals for the next election and therefore focus more on issues that are relevant to their local voters than on broader and more long-term challenges, such as climate change (Sweeney, personal communication, 23 November 2018). For many key actors in the political arena, climate change is an issue in the future and not something they can win elections with, which makes it less of a priority in their election campaigns and policies (Sweeney, personal communication, 23 November 2018).

As a consequence, one of the major reasons for Ireland being rather late in developing its health adaptation measures is that it was not able to reduce the high uncertainty that is connected with the topic by learning more about the concrete health risks the country is facing and what it can do about it. Although much more work has been done in this field recently, it will take some time until the results of the work will become visible.

The social and cultural dimension

Although Ireland's population is generally aware of climate change, the nexus between climate change and health is not a priority for many people (DCCAE, personal communication, 22 November 2018). At the same time, numerous civil society organizations have been established in the field of climate change, which from time to time address health issues as well (DCCAE, personal communication, 22 November 2018). They include Stop Climate Chaos, Friends of the Earth, IEN, An Taisce, and others (DCCAE, personal communication, 22 November 2018). Additionally, Ireland's Citizens Assembly brings together citizens and policymakers to discuss issues of national importance, and recently one of the key topics there was how to make Ireland an international leader in combating climate change (DCCAE, personal communication, 22 November 2018). Moreover, the National Dialogue on Climate Change seeks to gain feedback from the public on the topic and to get people more involved (DCCAE, personal communication, 22 November 2018). Although numerous NGOs exist, interviewed experts claimed that their influence was more limited than in other countries since they were seen very critically by public officials (Sweeney, personal communication, 23 November 2018). Even today, they are often not seen as partners and have limited possibilities to contribute to policymaking, especially when compared with private sector representatives (Sweeney, personal communication, 23 November 2018).

In the past, the climate change discourse was largely centered around mitigation, but adaptation has been gaining influence recently (DCCAE, personal communication, 22 November 2018). An expert stated, however, "[adaptation] will always be the second fiddle to mitigation" (DCCAE, personal communication, 22 November 2018). At the same time, another expert stated that many people in Ireland have a real attachment to land because of the colonial past during which people had no private property, which leads to strong resistance against sacrificing land for flood control measures (Sweeney, personal communication, 23 November 2018).

The economic dimension

As some societal barriers to adaptation suggest, the economy plays an important role for decisions on climate change in Ireland. Agriculture, especially the dairy industry, has a strong influence in Ireland and often opposes stronger commitments to climate action, which mostly affects the government's mitigation strategies (Institute of Public Health, personal communication, 22 November 2018; Technical University of Ireland, personal communication, 24 November 2018). Whilst the industry's interests with regards to adaptation are less clearly communicated, some interests against adaptation efforts exist as well (Sweeney, personal communication, 23 November 2018). An example is planned adaptation measures on flood control, which would involve dedicating more land to flood prevention, which can then not be used for agriculture (Sweeney, personal communication, 23 November 2018). Ireland's agriculture lobby is very strong and tries to prevent such measures whenever possible (Sweeney, personal communication, 23 November 2018).

The international dimension

Numerous experts have stated that the strongest international influence on adaptation to climate change in Ireland originates from the EU, which puts pressure on its Member States to effectively act against climate change and its consequences (Sweeney, personal communication, 23 November 2018; Technical University of Dublin, personal communication, 24 November 2018). When EU institutions criticize Ireland's work, the government often feels inclined to act (Technical University of Dublin, personal communication, 24 November 2018). Additionally, the EU plays an important role in providing standards, such as ISO standards, and mainstreaming and integrating policy processes on adaptation to climate change (DCCAE, personal communication, 22 November 2018). EU regulations require Ireland to take action on adaptation and report their progress to the EU (Department of Health, personal communication, 22 November 2018).

Health and the environment, however, are also at the EU level often treated as separate issues, and health is generally rather independent from the EU level since many states prefer to keep the main responsibility for health policies at the national level (Technical University of Dublin, personal communication, 24 November 2018). Furthermore, according to Irish experts, the current European Commission has been rather weak in putting pressure on governments, and it is still unclear whether the European Commission will enforce the targets set out in its 2020 strategy (Sweeney, personal communication, 23 November 2018). Therefore, as one expert put it, the EU's "scare factor" for the Irish government has recently been much lower than in the past (Sweeney, personal communication, 23 November 2018). Concurrently, politicians in Ireland also often use the EU as a scapegoat to sell unpopular actions to the public, which can be helpful for their climate policies as well (Sweeney, personal communication, 23 November 2018).

At the same time, a number of research projects with partners in the EU exists, especially in terms of adaptation to floods (DCCAE, personal communication, 22 November 2018). Moreover, numerous research projects have been initiated under the EU's Horizon 2020 strategy and funding for climate related topics has been given out (MaREI Centre, personal communication, 23 November 2018). The European Academy of Sciences is preparing a document on climate change and health for the European Commission, which is supposed to be published in 2019 and might lead to further momentum on the EU level (Technical University of Dublin, personal communication, 24 November 2018).

The UN and other organizations have significantly less direct influence on adaptation policies in Ireland since their focus is usually more on developing countries (Department of Health, personal communication, 22 November 2018; MaREI Centre, personal communication, 23 November 2018; Technical University of Dublin, personal communication, 24 November 2018). The largest indirect influence from international organizations stems from the IPCC through the publication of reports, such as the recent "IPCC Special Report on Global Warming of 1.5 °C" (DCCAE, personal communication, 22 November 2018). IPCC reports are often used by the Irish government to justify their actions, but none of the experts was aware of any direct cooperation between UN organizations and Ireland or policy learning from international institutions (Sweeney, personal communication, 23 November 2018).

Although not very widespread and influential, cooperation with other countries in the field of climate change adaptation exists. This is particularly true for cooperation with Northern Ireland, since under the Good Friday Agreement from 1998, Ireland cooperates with Northern Ireland on a number of policy areas, including the environment and climate change (DCCAE, personal communication, 22 November 2018). The Institute of Public Health is based on both sides of the Irish Border and some limited cooperation exists in terms of climate change adaptation. However, health adaptation is not a part of it (Institute of Public Health, personal communication, 22 November 2018). Under the British-Irish Council, environment ministers across the UK and Ireland meet and discuss environmental issues that concern both countries, yet climate change and health has not been much prominent on the agenda recently (DCCAE, personal communication, 22 November 2018). Cooperation with other European countries is rather low (DCCAE, personal communication, 22 November 2018).

Summary

Ireland's low CHAIn ranking can be explained by a combination of different factors, out of which the key variable is how the government has dealt with the uncertainty and complexity of the nexus between climate change and health. Due to lacking capacities in the Health Department and governmental agencies, as well as different urgent topics on the political agenda, such as the health sector crisis and the overall economic crisis from 2008 onwards, the country was not able to develop the groundwork-level measures early on to understand the specific impact of climate change on health in Ireland. Civil society organizations exist, but they have not put significant pressure on the government to strengthen health adaptation measures. With the Climate Action and Low Carbon Development Act from 2015 and subsequent processes of institutionalization and expanding adaptation measures, Ireland's health adaptation measures have started to develop and a major increase in initiatives is expected for the end of 2019.

Despite occasional strong influence by the EU, especially through reporting requirements from Ireland to the EU in terms of progress on adaptation, the international community has not significantly pushed Ireland to strengthen its adaptation work. International organizations, such as the WHO, only play a minor role. Among UN organizations, the IPCC is most respected, and its publications are broadly received by the Irish officials. Cooperation with other countries exists from time to time but has not led to significant improvements in Ireland's health adaptation work yet. Overall, Ireland's economic situation often takes priority for policymakers, and further institutionalization of groundwork- and adaptation-level actions on climate change and health will likely depend on the overall state of the Irish economy and the progress of interdisciplinary cooperation in Ireland's public agencies and research community.

Notes

1   Importantly, when this thesis was finalized, Ireland's adaptation work was undergoing lots of changes, especially with regard to the development of the new climate change adaptation framework and the sectoral adaptation plans that was published in 2019 (MaREI Centre, personal communication, 23 November 2018). As a new climate change and health adaptation plan exists now, it can be expected that Ireland will rank much higher on the CHAIn then before: https://www.dccae.gov.ie/documents/Health_Climate_Adaptation_Plan.pdf.
2   At this point, it is important to reemphasize that the number of initiatives does not necessarily reflect all measures on the ground since the index score depends on how comprehensively governments report on their adaptation initiatives in their national-level documents.
3   See EU summary report on zoonoses, zoonotic agents and food-borne outbreaks 2017 p. 92 at www.efsa.europa.eu/efsajournal.

References

Cullen, Elizabeth. 2007. *Climate Change and Health in Ireland: A national vulnerability assessment*. PhD thesis. National University of Ireland, Maynooth.

Department of Housing. 2019. "Water quality and water services infrastructure – climate change sectoral adaptation plan." Department of Housing, Planning and Local Government, Ireland.

Dinham, Paddy. 2017. "'Foolish and dangerous': video shows 'crazy' thrillseeker leaping 20ft into raging swell as Hurricane Ophelia batters Irish coast." *Daily Mail Online*, 16 October 2017. https://www.dailymail.co.uk/news/article-4986438/Irish-thrillseeker-leaps-sea-Hurricane-Ophelia.html.

ND-GAIN. 2019. "Ireland." Accessed 23 May 2019. https://gain-new.crc.nd.edu/country/ireland.

10 The Republic of Korea

Index test

The Republic of Korea (RoK) ranks at place 2 on the CHAIn and therefore represents one of the world's early adapters in terms of climate change related health risks. The expert interviews have shown that the RoK is indeed putting a lot of effort into its health adaptation initiatives and most of the adaptation actions that are outlined in the national adaptation strategies have been implemented in practice (KACCC, personal communication, 7 August 2018). Nevertheless, they come with a number of pitfalls. One example includes new infrastructure to adapt to heat-related health risks: although cooling centers for vulnerable populations exist, many people do not know where they are and how to access them (Seoul National University, personal communication, 6 August 2018). Moreover, some interviewed experts claimed that, due to its internal structures and working habits, the government and its agencies are very good at quickly developing strategies and communicating their actions, which sometimes leads to a better performance on paper than in practice (KEI, personal communication, 7 August 2018; Seoul National University, personal communication, 6 August 2018).

Nevertheless, the RoK is doing much more on health adaptation to climate change than the vast majority of other states. Additionally, although the majority of adaptation documents is published in English, some of the documents only exist in Korean and may contain additional health adaptation measures that were not included in the index (Seoul National University, personal communication, 6 August 2018). As a consequence, even if not all initiatives that make up the RoK's index score have been implemented, the overall number of measures is still much higher than in other countries (Seoul National University, personal communication, 6 August 2018). As a matter of fact, the Center for Climate Change Adaptation (KACCC) confirmed that the implementation rate of their earlier adaptation programs was very high, partly because many initiatives were summarized there that had existed for many years (KACCC, personal communication, 7 August 2018).

Governance of health adaptation to climate change in the RoK

International comparison

With 103.25 points, the RoK ranks at place 2 on the CHAIn, right after the UK (109), and is followed by Jordan (71.25), Canada (70.5), and Cyprus (61). Although the RoK's recognition score is relatively high (5.5 out of 7), it is slightly lower than that of many countries that rank high on the CHAIn, such as the UK (7), Jordan (7), and Canada (7), which shows that the RoK's is aware of fewer climate change related health risks than other countries. Moreover, with only 9.75 points, the RoK has a significantly lower groundwork score than the UK (42), Jordan (56.25), or Sierra Leone (16.5). The RoK's overall high CHAIn ranking can be explained by its exceptionally high adaptation-level score. With 88 points in the adaptation category, the RoK has the highest adaptation score of all analyzed countries. This shows that despite a lower amount of strategies and plans, policy recommendations, research, and communication tools, the RoK has implemented a great number of adaptation-level actions.

Health adaptation portfolio

The RoK's health adaptation portfolio is rather balanced with adaptation initiatives on all risk types, except for cold-related risks and tertiary climate change related health risks. With the groundwork-level score being comparatively low and the adaptation-level score exceptionally high, it clearly shows that heat-related health risks and infectious diseases are the major focus of the RoK. As Figure 10.1 indicates, the RoK has reported four groundwork initiatives on heat-related risks, four on infectious diseases, and three on risks that address the nexus between air pollution and climate change. Moreover, with two general groundwork-level measures, two on risks related to UV radiation, two on water and sanitation, and two on allergies, the East Asian country has a broad spectrum of groundwork-level initiatives. Additionally, the government reported one measure in the category floods and storms, one in the category droughts and fires, and one in the category food and agriculture.

The RoK's adaptation-level measures are less evenly distributed, with a clear dominance of initiatives that address heat-related health risks (14), infectious diseases (9), allergies (7), and general measures (5), shown in Figure 10.2. Three initiatives each were reported in the categories floods and storms and water and sanitation. For the categories droughts and fires, UV radiation, and food and agriculture, one adaptation initiative each was reported. The documents did not mention any initiatives on cold-related risks, risks related to air pollution, and tertiary health risks.

The overview of adaptation types in Figure 10.3 shows a clear dominance of adaptation actions that seek to change existing practices and behaviors (17

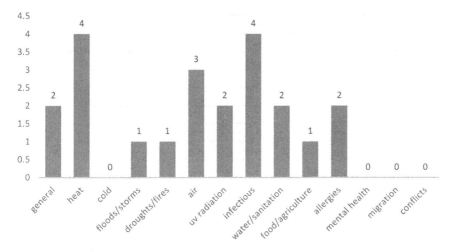

Figure 10.1 Distribution of groundwork-level initiatives in the RoK

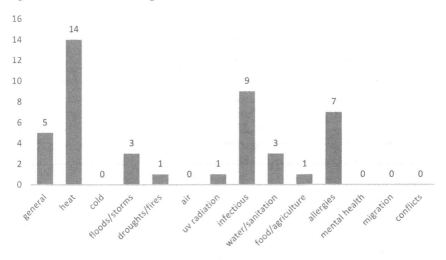

Figure 10.2 Distribution of adaptation-level initiatives in the RoK

measures), as well as research and development projects (13). Moreover, the RoK has a high number of monitoring and warning (9) and capacity building measures (7). Compared with other countries, the number of recommendations (5), strategies and plans (4), and communication tools (4) is low, but with five policies on health adaptation to climate change, the RoK is one of the top countries in that category. Additionally, the government reported four measures to educate people and raise awareness for climate change related health risks, one infrastructure measure, and one financing measure. Despite always being regarded as a frontrunner in technology innovations, the RoK did not report any health adaptation measures in the category technology.

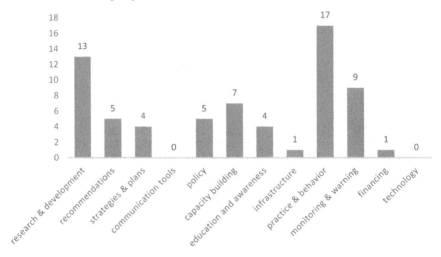

Figure 10.3 Distribution of adaptation types in the RoK

Overall, the portfolio shows that, while the RoK has a very strong focus on heat-related risks and infectious diseases and has implemented a great number of adaptation-level actions, especially in the categories practice and behavior and monitoring and warning, it does not mention any tertiary climate change related health risks, such as mental health risks and risks related to migration or conflicts. Despite being internationally known for its technological innovations, the RoK does not seem to make use of its leverage in this field by introducing technological innovations to reduce and prevent climate change related health risks.

Institutional framework and key actors

The Korea Environmental Institute (KEI), especially its Center for Climate Change Adaptation (KACCC), is essential for adaptation to climate change in the RoK (Seoul National University, personal communication, 6 August 2018). The KACCC was established to coordinate, integrate, and communicate the government's adaptation policies and has become the RoK's most important institution in the field (KACCC, personal communication, 7 August 2018; Seoul National University, personal communication, 6 August 2018). Whilst the respective ministries are responsible for developing sector-specific adaptation measures, which means that health adaptation primarily takes place in the Ministry of Health and Welfare, the KACCC has a strong influence on the processes and outcomes since it sets joint standards and contributes to evaluating the sector-specific work (KACCC, personal communication, 7 August 2018). Although the KEI, to which the KACCC belongs, reports directly to the Office of the Prime Minister, the KACCC reports to the Ministry of Environment and conducts research for the ministry (Seoul National University, personal communication, 6 August 2018). It works closely with several other research institutes and acts as a boundary organization between science and policy (KACCC, personal communication, 7 August 2018).

Changes over time

The development over time clearly shows a massive increase in adaptation-level initiatives with the government's official document on health adaptation to climate change from 2010 (Hae-Ryun et al. 2012). Reporting on health adaptation measures in the NCs to the UNFCCC is significantly lower, although an increase over time can be attested.

Without the specific document on climate change related health risks, however, the RoK would rank much lower on the CHAIn. As a consequence, it is worth delving into the drivers of and barriers to health adaptation to climate change in the RoK to gain a better understanding of why the number of adaptation-level initiatives is much higher than elsewhere.

Drivers and barriers

Climate change and health in the RoK – risk assessment

According to the ND-GAIN index, the RoK has a comparatively low vulnerability score (0.375) and ranks at place 16 on the overall ND-GAIN index (ND-GAIN 2019). Despite some vulnerabilities, ND-GAIN argues that the RoK is "well positioned to adapt" and represents the forty-seventh least

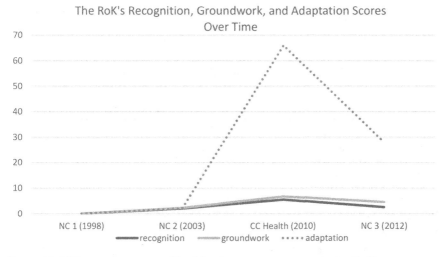

Figure 10.4 Time comparison of health adaptation measures in the RoK

vulnerable country in the world (ND-GAIN 2019). Since the ND-GAIN index does not solely focus on climate change and health, it is important to look at some health-specific risks to determine whether the general low vulnerability applies to the health sector as well.[1]

The index shows comparatively high scores when it comes to the "projected change of cereal yields" (0.822) and the "projected change of flood hazard" (0.835), which indicates that the RoK can expect an increase in flood-related health risks as well as risks related to food and agriculture if it does not adapt accordingly. Moreover, the RoK can expect a medium to high change in vector-borne diseases (ND-GAIN Score of 0.660), whereas the projected change of deaths from climate change induced diseases is comparatively low with an ND-GAIN score of 0.091. Similarly, despite the RoK's recent experience with heatwaves, the projected change of warm periods is lower than in other countries (ND-GAIN score 0.176).[2]

The projections of the World Bank's Climate Change Knowledge Portal show a higher probability of heatwave for the RoK than in other countries. With a projected probability of 0.16 under a medium low emission scenario (RCP 4.5) between 2040 and 2059 and 0.21 under a high emission scenario (RCP 8.5) for the same time period, the likelihood of severe heatwaves definitely increases for the RoK, so that the country can expect a similar increase in heat-related health risks as climate change progresses. Cold waves will become slightly less likely in the RoK. It shows that the projected change in annual probability of cold waves is −0.01 for 2040 to 2059, both under a medium low (RCP 4.5) and a high emission scenario (RCP 8.5). The change in annual severe drought likelihood is projected to slightly increase under a medium low emission scenario (RCP 4.5) to 0.06 for 2040–2059 and 0.05 under a high emission scenario (RCP 8.5). Compared with other countries, the projection for the RoK represents a rather low increase.

The risk assessment shows that the major climate change related health risks for the RoK are related to heat, floods and storms, and infectious diseases. The adaptation portfolio shows that the current focus of the government in terms of adaptation is on heat-related health risks and infectious diseases. This leads to the assumption that the government is well aware of

Table 10.1 ND-GAIN scores for climate change related health risks in the RoK

Risk	ND-GAIN score
Projected change of cereal yields	0.822
Projected change of deaths from climate change induced diseases	0.091
Projected change in vector-borne diseases	0.660
Projected change of warm periods	0.176
Projected change of flood hazard	0.835
Projected change of sea level rise impacts	0.160

the specific health risks climate change entails and has targeted them in its adaptation work. This is, however, not the case for risks related to floods and storms. Although the RoK is exposed to such risks, this area is not a priority in its health adaptation work yet. Consequently, the level of health adaptation cannot solely be explained by the actual exposure and vulnerability to such risks, but the risk perception and other factors are key to understanding what drives health adaptation in the RoK.

Risks assessment by experts from the RoK

Experts from the KEI claim that the most relevant climate change related health risks for the RoK are extreme heat, heavy rain, and risks related to food and agriculture (KEI, personal communication, 7 August 2018). Extreme weather events, especially typhoons, are already associated with climate change in the RoK and will become even more relevant in the future (KEI, personal communication, 7 August 2018). Additionally, infectious diseases, such as malaria, dengue fever, and others, will gain relevance as climate change progresses (Hae-Ryun et al. 2012). This goes along with the findings based on data from the ND-GAIN index and the World Bank. It is therefore worth taking a closer look at the perception of these risks and the political agenda in the RoK.

Risk perception and the political agenda

After years of exceptionally strong typhoons in the early 2000s, the Korean government dedicated more and more attention to climate change, initiating research projects on adaptation, and establishing the KACCC in 2009 (KEI, personal communication, 7 August 2018; Seoul National University, personal communication, 6 August 2018). Although, compared with other countries, media coverage on the health effects of climate change and climate change adaptation is relatively low, when extreme weather events happen the topic receives more attention and makes it to the top of the political agenda (Seoul National University, personal communication, 6 August 2018). In 2018, a number of typhoons and a strong heatwave demonstrated the impact of climate change again and the topic was largely covered by the media, but it did not trigger a significant change in the government's policies in the field (KACCC, personal communication, 7 August 2018). Overall, climate change adaptation has been on the government's radar for a long time, especially since the country is frequently hit by extreme weather events, but it is not a key priority for the government at the moment (KACCC, personal communication, 7 August 2018).

The country's high CHAIn score can to a great extent be explained by the strong institutions on climate change adaptation. The most important pillar of the RoK's adaptation work was established as a consequence of extreme weather events in the early 2000s, which were key to setting the topic on the political agenda. Due to the government's perception of these formative events and its effective response by setting up new institutions to tackle the risks, the country is much better prepared to adapt than other countries across the globe, which may have experienced similar events. Consequently, how states perceive climate change and how they learn from formative events is an essential factor that drives their adaptation efforts.

Dealing with complex decisions under uncertainty

Since the Korean government recognized that adaptation to climate change is a very complex issue that requires the capacity to deal with long-term challenges and levels of uncertainty that regular governmental institutions do not possess in their day-to-day activities, it decided to establish the KACCC (KACCC, personal communication, 7 August 2018). The KACCC provides the required expertise and network to foster research on climate change adaptation and develop appropriate actions (KACCC, personal communication, 7 August 2018). As climate change is a very inter-disciplinary issue, one of the KACCC's priorities is to connect researchers and policymakers from a variety of backgrounds and develop integrated solutions (KACCC, personal communication, 7 August 2018). The KACCC streamlines the government's policies in the field, conducts its own research, brings together key stakeholders, and overall is the most important success factor for the RoK's adaptation initiatives. Moreover, the KEI and the KACCC cooperate closely with public and private research institutes and universities, such the Samsung National Disaster Institute (KEI, personal communication, 7 August 2018). They hold regular meetings together and share their experiences to learn from each other and improve their own adaptation initiatives to climate change (KEI, personal communication, 7 August 2018). As a consequence, the RoK is well equipped to deal with the complexity and uncertainty that health adaptation to climate change entails. Nevertheless, high boundaries between the different ministries and research institutes still remain and more cross-sectoral and interdisciplinary cooperation is necessary (KACCC, personal communication, 7 August 2018).

The social and cultural dimension

The Korean public is generally aware of climate change and the topic is frequently covered in the media (Seoul National University, personal communication, 6 August 2018). However, some interviewed experts claimed that the news on the topic is often the same and the public gets bored and disconnected from it (Seoul National University, personal communication, 6 August 2018). Additionally, large parts of the public discourse are dominated by mitigation and energy efficiency, whereas adaptation only plays a very minor role (Seoul National University, personal communication, 6 August 2018). Moreover, many people believe that climate change is something happening far away and not in Korea (Seoul National University, personal communication, 6 August 2018). As one expert put it, from the perspective of the Korean population, climate change is a "polar bear issue" and not something that effects them personally (Seoul National University, personal communication, 6 August 2018). Although civil society organizations such as the Korea Federation for Environmental Movements exist, they do not play a key role in the public discourse on climate change adaptation. As a consequence, the society's pressure on the government to act more is comparatively low and cannot explain the RoK's high ranking.

The economic dimension

The economy plays an outstanding role in South Korean politics and is very frequently on top of the political agenda (Seoul National University, personal communication, 6 August 2018). As a very export-oriented country, the RoK takes its reputation in the international arena very seriously and is therefore very receptive to the discussions and outcomes of international climate change negotiations, such as the Conferences of the Parties to the UNFCCC (Seoul National University, personal communication, 6 August 2018). Moreover, the RoK sees itself as a bridge between developing and developed countries, especially when it comes to mitigation goals (Seoul National University, personal communication, 6 August 2018). Compared with developing countries, the RoK pursues an international leadership role in terms of climate change, but at the same time it believes it has fewer responsibilities than most developed countries (Seoul National University, personal communication, 6 August 2018). Whilst the RoK's pursuit of international reputation to foster its export-oriented economy definitely constitutes a significant factor, it cannot solely explain the RoK's strong adaptation portfolio.

The international dimension

Through its main institution in the field of climate change adaptation, the KACCC, the RoK cooperates frequently with international organizations, such as the IPCC and the UNFCCC, especially with regards to the international climate change adaptation expo in South Korea in 2019 (KACCC, personal communication, 7 August 2018). Although the RoK does not directly benefit from financial or technical support from international organizations in the field of climate change adaptation, it cooperates frequently with UN organizations to foster knowledge and technology transfer (KACCC, personal communication, 7 August 2018). Through UN Environment and UNDP, the RoK provides support and advice to developing countries that seek to improve their adaptation plans and measures (KACCC, personal communication, 7 August 2018). The RoK often hosts international events on adaptation, such as the Regional NAP Expo in 2017, and fosters international transfer of knowledge and expertise on climate change adaptation (NAP-EXPO 2017). Additionally, the RoK cooperates with a number of developed states, especially Germany, Japan, and the UK, to share knowledge about adaptation to climate change and learn from each other (KACCC, personal communication, 7 August 2018). The KACCC was, for instance, involved in the KOMPASS adaptation program of the German government and shared information on adaptation processes and experiences in the RoK (KACCC, personal communication, 7 August 2018).

On a broader level, the international community plays an important role as a norm and agenda setter for the RoK. As one expert put it, the RoK "is a follower, not an initiator" in international climate change negotiations and thus takes international framework documents and guidelines very seriously (Seoul National University, personal communication, 6 August 2018). Due to the fact that recent international climate change conferences and agreements, such as the Paris Agreement from 2015, place more emphasis on climate change adaptation, the topic has gained relevance in the RoK as well (Seoul National University, personal communication, 6 August 2018). Moreover, publications of the IPCC are closely watched and often spark new debate on the topic in the RoK (Seoul National University, personal communication, 6 August 2018). Key institutions cite the IPCC's work very frequently and build their own research and adaptation efforts on the IPCC's findings (KEI, personal communication, 7 August 2018; Seoul National University, personal communication, 6 August 2018).

Summary

The RoK is a global leader in health adaptation to climate change. Although the East Asian country does have slightly fewer implemented measures on the ground than its strategies and plans suggest, it still ranks much higher than almost every other country. The RoK's focus on heat-related health risks and infectious diseases mirrors the actual climate change related health risks that the country is confronted with. However, based on the risk assessment, it needs to work more to protect its population against health risks related to floods and storms. Overall, the RoK's adaptation measures have a very strong research background and the vast majority of its adaptation-level measures belong to the category "practice & behavior," which shows that the academic findings and strategies often lead to behavioral changes in the field. Interestingly, despite being renowned for technological innovation, the RoK did not report any adaptation measures in this category, which suggests that the composition of health adaptation measures really depends on what the key policymakers interpret as adaptation. Moreover, a significant difference between the reporting in NCs and specific health adaptation documents exist.

The most important reasons for the RoK's diverse health adaptation portfolio are its strong institutions that develop, coordinate, and integrate strategies, plans, and concrete measures on the subject. These institutions were mostly established after the RoK experienced a number of extreme weather events in the early 2000s, attributed them to climate change and learned that it had to increase its adaptation initiatives. The RoK's central institution on climate change adaptation, the KACCC, is the number one reason for the country's high performance since it acts as a boundary organization between research and policy and therefore enables the country to effectively solve complex policy challenges that involve high levels of uncertainty. The KACCC allows the Korean government to deal with long-term challenges that the government does not have the capacity to work on in its day-to-day activities and therefore ensures that it is better prepared for the future than other countries. Nevertheless, in order to further enhance its health adaptation work, even more cross-sectoral cooperation and interdisciplinary research is necessary in the RoK.

Civil society does not constitute a key driver or barrier in terms of health adaptation to climate change. The economy, on the other hand, is very export oriented, which is why the RoK depends on good international reputation and therefore strives to be seen as a global leader in terms of climate action. As a consequence, it closely follows international publications, especially those from the IPCC, climate change conferences, and other events in the field. It cooperates frequently with UN organizations and even helps developing countries to strengthen their adaptation plans and strategies. As a consequence, the RoK's political agenda is definitely influenced by the international community, and it constantly learns about current international adaptation trends that could benefit its own adaptation measures. Overall, the international community's role can be understood as a norm and agenda setter.

Notes

1 Although the ND-GAIN index contains a category on health, the risks and proxies under that category only represent a small fragment of the actual climate change related health risks countries have to face. Therefore, additional indicators from the general index are taken into account.
2 It needs to be noted, however, that this score only shows the projected change in this field and not the absolute numbers. Since the RoK is already experiencing high temperatures in summers, even a small increase can lead to severe health risks.

References

Hae-Ryun, Choi, Soonam Jo, Min-kyung, and Tae Won Son. 2012. *Climate Change and Human Health: Impact and Adaptation Strategies*. Korea: Ministry of Health and Welfare and Korea Centre for Disease Control & Prevention.
NAP-EXPO. 2017. "Regional NAP Expo." Accessed 2 June 2018. http://napexpo.org/asia/.
Notre Dame, Global Adaptation Initiative (ND-GAIN). 2019. "Republic of Korea." Accessed 26 February 2019. https://gain.nd.edu/our-work/country-index/.

11 Japan

Index test

The interviews with Japan's most renowned scientists and policymakers on health adaptation to climate change have helped to identify the most important documents in the field. They include the NCs to the UNFCCC, the National Plan for Adaptation to Climate Change, the Report on Assessment of Impacts of Climate Change in Japan and Future Challenges of the Central Environmental Council, and a small number of additional documents, which are only available in Japanese and not in English (MiLai, personal communication, 31 July 2018; Ueda and Xerxes, personal communication, 2 August 2018). Most of the documents that experts mentioned were already included in the database of the CHAIn and the ones that were not identified earlier do not include a significant amount of additional measures that the country has undertaken to respond to climate change related health risks. Consequently, Japan's ranking on the CHAIn can generally be confirmed for the analysis period. In June 2018, however, the government adopted the new Climate Change Adaptation Act, which seeks to promote adaptation measures in various fields, including human health (Japan, Ministry of Environment 2018a). The Comprehensive Adaptation Programme under the Climate Change Adaptation Act seeks to establish clearer roles of national and local governments and other key stakeholders in terms of who is responsible for adaptation and who can take which actions. Moreover, it requires the national government to adopt a new National Adaptation Plan (NAP) to promote adaptation in "all sectors" and to develop methods to monitor and evaluate the progress of the different sectors (Japan, Ministry of Environment 2018b). Last but not least, the act asks the Ministry of the Environment to conduct climate change impact assessments on a five-yearly basis (MiLai, personal communication, 31 July 2018).[1] Compared with other countries, such as the RoK, the UK, or Germany, Japan is, however, rather late with stepping up their adaptation action, especially in terms of climate change and health. Despite growing awareness of the impact of climate change in Japan, the interviews have shown that health adaptation is not a key priority yet, partly because mitigation has for a long time taken precedence over adaptation, as the next paragraphs will show.

Governance

International comparison

With an overall score of 13.75 points, Japan ranks at place 85 on the CHAIn index, just above San Marino (13.5 points) and Saint Lucia (13.25 points) and below Guatemala (13 points). It shares rank 85 with the Democratic People's Republic of Korea and the Islamic Republic of Iran, which both have a score of 13.75 points as well. Japan has a comparatively high recognition score (4.5 out of 7), which indicates that the country recognizes a significant number of climate change related health risks. At the same time, its groundwork-level and adaptation-level initiatives are very low, as the following section will demonstrate. Compared with other countries with similar exposure to climate change and similar economic and political determinants, such as the RoK, Japan clearly underperforms on the index.

Health adaptation portfolio

Japan's adaptation documents show a clear dominance of groundwork-level initiatives that address heat-related health risks (12) and infectious diseases (5). Two general measures and two initiatives from the category "water and sanitation" were reported. Only one measure each in the categories "floods and storms," "droughts and fires," and "air pollution" were reported. The documents did not contain any initiatives addressing cold-related health risks, UV radiation, food and agriculture, allergies, or any tertiary climate change related health risks. This shows that, despite a high recognition score, Japan's groundwork-level measures have a very strong focus on heat-related risks and ignore the majority of other climate change related health risks.

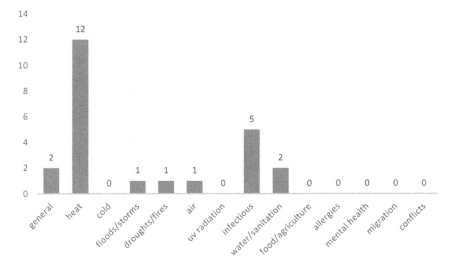

Figure 11.1 Distribution of groundwork-level initiatives in Japan

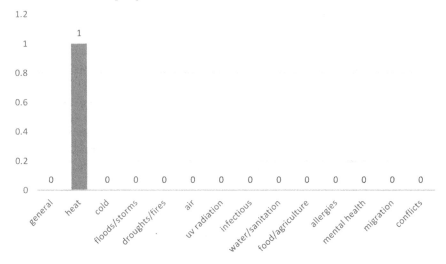

Figure 11.2 Distribution of adaptation-level initiatives in Japan

Overall, Japan's adaptation portfolio mostly consists of groundwork-level measures and only one heat-related adaptation-level initiative was reported.

As Figure 11.3 illustrates, Japan's health adaptation portfolio is heavily dominated by recommendations (19), which shows that the state is aware that certain actions are necessary and concrete proposals are made on what should be done, but not much action has followed from these recommendations. With two measures in the category "strategies and plans" and two communication tools, as well as one reported measure in the category "research and development" and one in the category "infrastructure"

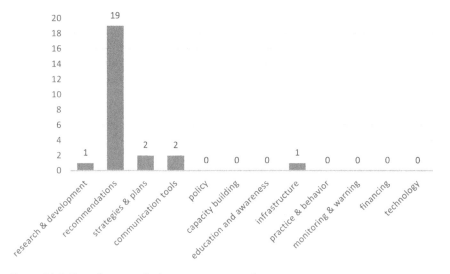

Figure 11.3 Distribution of adaptation types in Japan

Japan's adaptation measures are overall quite narrow and do not include a high number of implemented actions.

The East-Asian country completely leaves out tertiary climate change related health risks in all adaptation-level categories (recognition, groundwork, and adaptation) and does not report any concrete national-level policies or programs to adapt to climate change related health risks.

Institutional framework and key actors

The key players for health adaptation to climate change in Japan are the Ministry of Health, Labour and Welfare, the Ministry for the Environment, and several governmental agencies and research groups. The National Institute of Environmental Studies (NIES), which is funded by the government and serves under the Ministry for the Environment, is an important institution in the field of climate change adaptation in Japan since it coordinates and contributes to a number of research projects in the field (Ueda and Xerxes, personal communication, 2 August 2018). The Japanese government has initiated several research projects under the NIES on climate change adaptation and public awareness of climate change (Ueda and Xerxes, personal communication, 2 August 2018). Moreover, the Environmental Restoration and Conservation Agency (ERCA) funds a variety of research projects on adaptation (Ueda and Xerxes, personal communication, 2 August 2018). A large part of Japan's research on climate change adaptation takes place under the Strategic Research on Global Mitigation and Local Adaptation to Climate Change (MiLai) project, which was established in 2014 under the Japanese Ministry for the Environment (MiLai, personal communication, 31 July 2018). The outcomes of the project are regularly presented in official reports and consulted by the Ministry for the Environment (MiLai, personal communication, 31 July 2018). Moreover, members of the Ministry for the Environment participate in the project meetings and often seek advice from the project group's key scientists (MiLai, personal communication, 31 July 2018).

Changes over time

The longitudinal assessment of Japan's adaptation policies from the first NC to the UNFCCC to the most recent NC in 2017 shows that, despite early recognition of the effects of climate change on health in the 1990s, it took the Japanese government until the adoption of the National Adaptation Plan in 2015 to develop a significant number of groundwork actions and until the publication of its seventh NC in 2017 to implement its first ever health adaptation actions. As Figure 11.4 illustrates, Japan was quite early in recognizing climate change related health risks in its international publications on climate change. The number of recognized risks and groundwork actions dropped in 2009 and then rose again after 2014.

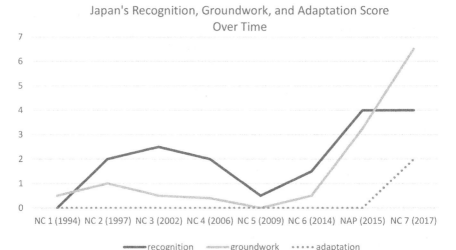

Figure 11.4 Time comparison of health adaptation measures in Japan

In international comparison, however, Japan's groundwork and adaptation actions are still on a comparatively low level. With a recognition score of 4, Japan acknowledges more than half of the climate change related health risks that international academia has identified. As of 2018, the Japanese government has undertaken many efforts to improve its adaptation performance. As a consequence, Japan's scores will likely rise in the near future.

Drivers and barriers

Climate change and health in Japan – risk assessment

With a generally low vulnerability score (0.372) and a high readiness score (0.674), Japan is the forty-third least vulnerable country and the seventeenth most ready country on the ND-GAIN index (ND-GAIN 2019). The health-specific data from the ND-GAIN index suggests that the major climate change related health risks for Japan are related to food and agriculture, which is reflected in a high score for the projected change of cereal yields (0.798) and extreme weather events, such as floods (projected change of flood hazard is 0.763). Moreover, Japan can expect a significant increase in the risk of infectious diseases since the projected change in vector-borne diseases has a relatively high ND-GAIN score (0.658). The projected change of deaths from climate change related diseases, however, is 0, which can be explained by Japan's good health system and high readiness to adapt to climate change related health risks. Heat-related health risks constitute a challenge, but with a ND-GAIN score of 0.365 in the projected change of warm periods, other health risks take precedence.

Table 11.1 ND-GAIN scores for climate change related health risks in Japan

Risk	ND-GAIN score
Projected change of cereal yields	0.798
Projected change of deaths from climate change induced diseases	0
Projected change in vector-borne diseases	0.658
Projected change of warm periods	0.365
Projected change of flood hazard	0.763
Projected change of sea level rise impacts	0.350

The data from the World Bank's Climate Change Knowledge Portal generally confirms the ND-GAIN findings. With a projected probability of 0.08 under a medium low emission scenario (RCP 4.5) between 2040 and 2059 and 0.10 under a high emission scenario (RCP 8.5) for the same time period, the likelihood of severe heatwaves will increase, but not as much as in other countries. The likelihood of cold waves will decrease by −0.01 between 2040 and 2059 both under a medium low (RCP 4.5) and a high emission scenario (RCP 8.5). After 2059, the annual probability of a cold wave is projected to decrease even further to −0.02. Similarly, severe droughts will become slightly less likely in Japan. Compared with other countries, however, the projected change in annual severe drought likelihood is rather moderate. Under a medium low emission scenario (RCP 4.5), it is projected to increase to 0.06 for 2040–2059, and to 0.03 under a high emission scenario (RCP 8.5).

To summarize, the most relevant health risks that are associated with climate change in Japan are related to infectious diseases, floods and storms, as well as food and agriculture. This shows a clear discrepancy between the actual risks and the health adaptation measures since most of Japan's adaptation initiatives focus on heat-related health risks. Although heat-related health risks should not be underestimated, the gap between the taken measures and actual risks shows that risk perception and attribution is a much stronger driver of health adaptation to climate change than actual risks. Japan has been exposed to severe heat in recent years, which may have led to more measures in this field, although the risk assessment shows that other risks, such as floods and storms, require even more measures at the moment.

Risks assessment by experts from Japan

According to Japanese experts on climate change adaptation, the major climate change related health risks for the country are those related to extreme weather events, such as storms, floods, and heatwaves (MiLai, personal communication, 31 July 2018; Ueda and Xerxes, personal communication, 2 August 2018). Tertiary climate change related health risks, such as mental health risks or risks related to migration and conflicts, were not mentioned by any expert and are also not discussed in official documents on climate change and health. Overall, the risk assessment by Japanese experts goes along with the evidence gained from the ND-GAIN index and the World Bank's Climate Change Knowledge Portal and shows that extreme weather events are the number one climate change related health risk in Japan.

Risk perception and the political agenda

The nexus between climate change and health is not a priority for the Japanese government. Other topics within the climate change debate rank higher on the government's agenda (MiLai, personal communication, 31 July 2018). Especially after the Fukushima Daiichi nuclear disaster in 2011, energy security was at least partially prioritized over climate change adaptation, and the government frequently made public statements on the importance of energy efficiency and reducing energy consumption, even in summer (MiLai, personal communication, 31 July 2018). In the subsequent years, the government refrained from such statements since many people had suffered from heat exhaustion and heat stroke because they did not turn on their air conditioning in the summer to save energy (Kobe City University of Foreign Studies, personal communication, 16 June 2019). Moreover, although the Japanese society and government are generally aware of climate change, not much knowledge exists on the actual health effects of climate change on individuals (MiLai, personal communication, 31 July 2018). With the severe heatwave and numerous strong typhoons in 2018, climate change adaptation has gained importance on the political agenda in Japan, which coincides with the government's efforts to improve the country's national adaptation strategy (MiLai, personal communication, 31 July 2018).

Dealing with complex decisions under uncertainty

To make informed decisions on health adaptation to climate change, it requires a strong research basis on the expected effects of climate change on health in Japan. Although academia has recently started to do more research on the subject, and health plays an increasingly important role in national adaptation projects, such as the MiLai project, Japan's research community on climate change and health is still comparatively small. Although some internationally renowned experts, such as Prof. Dr. Yasushi Honda, have contributed largely to a better understanding of the specific health risks of climate change in Japan, the number of experts and research projects on the matter in other countries is much higher. Moreover, many of the existing research projects on the subject in Japan focus mostly on sub-topics that are particularly high on the agenda, such as heat-related health risks or risks associated with air pollution. In addition, the focus is much more on mitigation of climate change and the co-benefits of mitigation than on adaptation

(MiLai, personal communication, 31 July 2018). Similarly, the government's capacity to conduct research on the topic has only recently started to develop, especially when it comes to the Ministry for the Environment (MiLai, personal communication, 31 July 2018).

One of the major challenges in terms of health adaptation to climate change in Japan is policy integration (Ueda and Xerxes, personal communication, 2 August 2018). Health issues are the responsibility of the Ministry of Health and Welfare, while everything related to climate change is under the mandate of the Ministry for the Environment (Ueda and Xerxes, personal communication, 2 August 2018). However, the Ministry for the Environment's emphasis on the nexus between climate change and health is rather low since their focus is much broader and environment-focused (Ueda and Xerxes, personal communication, 2 August 2018). Despite some cooperation on the federal level, much more could be done on health adaptation and, especially on the level of prefectures, more cooperation between the different decision-making bodies in the fields of health and the environment is needed (Ueda and Xerxes, personal communication, 2 August 2018).

The social and cultural dimension

The government's focus on mitigation in recent years can be explained by the cultural and social perception of the term adaptation in Japan. Numerous experts stated that adaptation is not socially desirable in Japan because the society thinks that mitigation to climate change should be the key priority for the government and a focus on adaptation could endanger mitigation efforts (WHO, personal communication, 4 August 2018). As a matter of fact, the word "adaptation" is avoided and terms such as "eco" and "green" are utilized in governmental policies and discussions as well as by the private sector (WHO, personal communication, 4 August 2018). As a consequence, most of the government's policies on climate change and health are directed towards mitigation (Ueda and Xerxes, personal communication, 2 August 2018).

At the same time, due to numerous severe weather events, such as floods and storms, in recent years, and media coverage of these instances, Japan's society is generally very aware of climate change (Ueda and Xerxes, personal communication, 2 August 2018). Until very recently, however, large parts of the population shared the belief that climate change exists mostly outside of Japan (Ueda and Xerxes, personal communication, 2 August 2018). Additionally, the nexus between climate change and health used to be omitted by the media and politics.

The year 2018, however, led to a change of perceptions. Following the massive heatwave and numerous typhoons, heavy rains, and other extreme weather events, media coverage on the effects of climate change increased and, according to the research team, the public's perception began to change as well (Ueda and Xerxes, personal communication, 2 August 2018).

Additionally, with the government's adoption of the Climate Change Adaptation Act from 2018, the topic has since gained much more attention both in politics and the general public (Ueda and Xerxes, personal communication, 2 August 2018).

The economic dimension

As an economically strong country with a well-functioning health system and a very healthy society, in theory, Japan has the necessary resources to be among the world's leading nations in terms of health adaptation to climate change. In practice, however, Japan is lagging behind on health adaptation, especially when compared with its neighbor South Korea. Whilst economic capacities cannot adequately explain Japan's rank, the fact that other topics, such as the overall health of the economy, are higher on the political agenda may have more explanatory power. The economy plays an outstanding role for Japan's society and politics and therefore often dominates the political agenda (WHO, personal communication, 4 August 2018). The political discourse on health or environmental topics often contains many economic components and decisions are frequently made on cost–benefit calculations (MiLai, personal communication, 31 July 2018; WHO, personal communication, 4 August 2018). Compared with other topics, such as disaster risk reduction, climate change and health is more abstract and the economic losses associated with the topic are less clear, which is one of the reasons why the topic is less prominent on the agenda (WHO, personal communication, 4 August 2018). Where the government sees a clear link to the economy, as for instance in the case of natural disasters that may drastically affect the population and the economy, it is willing to invest many resources to protect the country against such risks (WHO, personal communication, 4 August 2018). In terms of climate change and health, the economic benefits of adaptation and the cost–benefit calculation, which plays an important role in all of Japan's adaptation discussions, are less clear and more research is therefore needed.

The international dimension

Japan's strong economy and health system lead to more independence and less exchange with the international community, since international organizations, such as the WHO, focus more on developing countries that need more support (WHO, personal communication, 4 August 2018). Due to Japan's high economic performance and good health system, the influence of international organizations on the country's health policies is rather low, which is why fewer concrete projects in cooperation with international organizations, such as the WHO, exist in the field of climate change and health (MiLai, personal communication, 31 July 2018; Ueda and Xerxes, personal communication, 2 August 2018). Although the WHO and other international organizations are active in Japan, there is no specific project or task force on climate change and health at the moment. Until 2011, a technical officer of the WHO in Kobe, Japan, specifically focused on climate change and health, but since then the focus has shifted more towards disaster risk reduction (WHO, personal communication, 4 August 2018). No specific cooperation projects in the field of climate change and health exist on the governmental level (MiLai, personal communication, 31 July 2018).

However, the IPCC and other key actors in the field, such as the WHO and UNFCCC, have a significant impact on the research community and the framing of the overall political discourse on climate change in Japan (MiLai, personal communication, 31 July 2018). The most recent IPCC reports and other publications of UN organizations are frequently cited by Japanese experts in the field (MiLai, personal communication, 31 July 2018). The MiLai project, one of the key research projects on climate change adaptation in Japan, followed the publication of the IPCC's fifth Assessment Report (MiLAi n.d.).

Summary

In international comparison, Japan is late at developing its health adaptation initiatives. Although the Japanese government has recently started to improve and update its adaptation framework, the nexus between climate change and health is not a key priority for the East Asian country. Most of its existing measures are groundwork-level initiatives that help to prepare future adaptation-level actions. Although Japan's new adaptation framework may contain additional health adaptation measures that could increase the country's CHAIn ranking, Japan still needs to strengthen its efforts to adequately respond to the climate change related health risks the country is confronted with.

Japan's relatively low ranking can be explained by the government's and public's perception of climate change related health risks and adaptation in general. Despite a number of severe weather events that can be attributed to climate change, neither the government nor the population took any direct actions to counter climate change related health risks as a consequence of those events. When thinking about climate change, many people had the impression that climate change was something that happened outside of Japan. With the severe heatwave and the numerous typhoons in 2018, this perception did, however, start to change and the pressure on the government started to rise. Moreover, the term adaptation used to have a negative connotation in the past since parts of the population feared that strengthened adaptation efforts would prevent further mitigation of climate change. The analysis has shown that risk perception and learning from formative events are a key factor in terms of health adaptation due to the dissonance between actual risks and addressed risks in the adaptation portfolio. Although the climate change projections show that Japan should strengthen its measures on floods and storms and infectious diseases, the most addressed risk is heatwaves. Heat-related health risks definitely constitute an important risk, but others should be addressed even more.

Additionally, despite its strong economy, Japan is lacking driving forces within academia and governmental institutions to strengthen its health adaptation work. Its research community in the field of climate change and health is not large enough and governmental institutions do not integrate policies effectively. Although the ministries and governmental agencies, as well as specific adaptation projects, such as MiLai, address climate change adaptation and do a lot of groundwork-level work, health adaptation is not a priority yet. Much of the work is still done within the specific sectors.

Moreover, due to its strong economy and independence, Japan does not get any direct support from international organizations in the field of health adaptation, although UN organizations, such as the WHO, have established offices in the country. The international community only indirectly influences Japan's political agenda and research on climate change, especially through IPCC publications. The nexus between climate change and health only recently started to make it to the top of the UN's agenda itself. Overall, the number major reasons for Japan's late and slow start with regards to health adaptation to climate change are its risk perception and lacking institutional drivers of strengthened actions in the country.

Note

1 As a consequence, if the adaptation assessment were to be conducted for Japan again, it is very likely that the country would rank much higher on the CHAIn.

References

Japan, Ministry of Environment. 2018a. *Climate Change Adaptation Act.* Japan: Ministry of Environment Japan.

Japan, Ministry of Environment. 2018b. *Climate Change Adaptation Act Overview.* Japan: Ministry of Environment Japan.

MiLAi. n.d. "Strategic R&D area project 'S-14' of the Environment Research and Technology Development Fund [MiLAi]." Accessed 26 February 2019. http://s-14.iis.u-tokyo.ac.jp/eng/overview/.

ND-GAIN. 2019. "Japan." Accessed 23 May 2019. https://gain.nd.edu/our-work/country-index/.

12 Sri Lanka

Index test

Sri Lanka's exceptionally high amount of health adaptation measures to climate change and its remarkable performance on the CHAIn raise a number of questions that this chapter seeks to answer: Is Sri Lanka's CHAIn score justified and do the measures on paper represent the actual level of health adaptation to climate change in the country? What are the drivers of and barriers to health adaptation in the South Asian country?

Sri Lanka's high CHAIn score does reflect the current state of health adaptation to climate change on paper but not always on the ground. Many interviewed experts claimed that Sri Lanka was very good at strategizing and communicating its efforts, partly also due to international support in this area through external consultants, but what is on paper is not always implemented on the ground (Amerasinghe, personal communication, 12 February 2018; Athukorala, personal communication, 12 February 2019; Journalist, personal communication, 13 February 2018). When discussing Sri Lanka's health adaptation policies, some experts spoke of a "paradise on paper" that is not always followed up with concrete actions (Athukorala, personal communication, 12 February 2019). This observation is confirmed by the fact that health adaptation to climate change is not part of Sri Lanka's new budget (Ministry of Finance, personal communication, 13 February 2019). As a consequence, funding for health adaptation needs to either come from other sources, such as international organizations, or the measures that are planned on paper are not implemented (Ministry of Finance, personal communication, 13 February 2019). Nevertheless, climate change and health is discussed and addressed significantly more often than in other countries and, based on the existing strategies and plans, Sri Lanka deserves its high CHAIn score.

Governance

International comparison

With rank 10 on the CHAIn, Sri Lanka belongs to the early adapters and global leaders in health adaptation to climate change. With a CHAIn score of 46.25, it ranks right after the Seychelles, another high-ranking island state, which also has a score of 46.25 but more groundwork-level initiatives than Sri Lanka. Sri Lanka ranks only two places below Germany (rank 8 with 47.5 points) and two ranks higher than the United States (rank 12 with 42 points). The South Asian island has a relatively high recognition score, which indicates that the country is aware of a broad bandwidth of different climate change related health risks. With a recognition score of 5.5 (out of 7 possible points), it has the same recognition score as the RoK, which ranks second on the overall CHAIn. Sri Lanka's medium high groundwork score (12.75) and high adaptation score (28) show a more balanced portfolio in terms of adaptation-levels than many other high-ranking states, such as the United States or Turkey, which both have high adaptation scores and low groundwork scores. Overall, Sri Lanka scores surprisingly high on the CHAIn and its sub-indexes, and therefore it is definitely worth further analysis to understand the drivers of and barriers to health adaptation in this South Asian country.

Health adaptation portfolio

With a recognition score of 5.5, Sri Lanka acknowledges the impact of climate change on a broad range of different health risks and, unlike many other high-ranking countries, even recognizes some tertiary climate change related health risks (mental health risks and those related to migration). At the same time, in its official documents, Sri Lanka does not recognize cold-related health risks, risks related to UV radiation, or risks related to conflicts.

Sri Lanka's groundwork-level initiatives heavily focus on infectious diseases (11 initiatives), risks related to air pollution (9 initiatives), and risks related to food and agriculture (7 initiatives) (Figure 12.1). Tertiary climate change related health risks or risks related to allergies, UV radiation, droughts and fires, or cold-related health risks are not mentioned in Sri Lanka's official documents.

The adaptation-level portfolio shows a different distribution (Figure 12.2). Unlike in the groundwork-level portfolio, Sri Lanka does not report any adaptation-level initiatives that address risks related to air pollution or heat-related health risks and mentions only one measure that addresses risks related to water and sanitation. Moreover, with three general measures, three measures on floods and storms, and three measures that address infectious diseases, the high number of groundwork-level measures on infectious diseases (11 initiatives) is not completely mirrored in the adaptation-level measures. The most reported adaptation-level initiatives address risks related to food and agriculture (4 initiatives). Tertiary climate change related health risks were not mentioned at all.

The distribution of Sri Lanka's adaptation types shows a relatively large number of recommendations (33). Moreover, the country reports

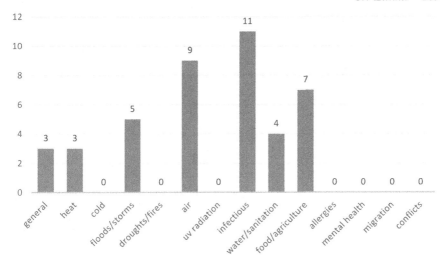

Figure 12.1 Distribution of groundwork-level initiatives in Sri Lanka

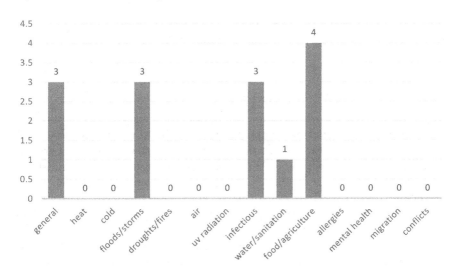

Figure 12.2 Distribution of adaptation-level initiatives in Sri Lanka

four research and development projects on climate change and health and five measures under the category strategies and plans (Figure 12.3). The high number of recommendations shows an acute awareness of the necessity to act and an understanding of what can be done, but it does not imply any implemented adaptation-level action. At the same time, however, Sri Lanka reports nine capacity building measures, which is a very high number in international comparison. This shows that Sri Lanka has implemented a relatively high number of concrete adaptation-level measures. Compared with the high number of recommendations, it

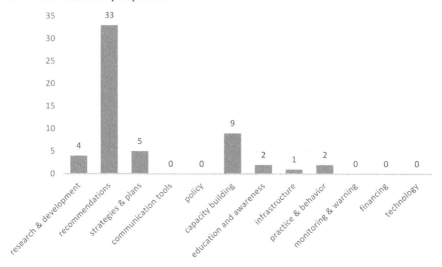

Figure 12.3 Distribution of adaptation types in Sri Lanka

becomes clear, however, that much more needs to be done on implementing the recommendations and turning them into concrete actions on the ground.

Institutional framework and key actors

Sri Lanka's central institution for adaptation to climate change is the Climate Change Secretariat under the Ministry of Environment and Mahaweli Development and sector specific departments (Ministry of Health, personal communication, 13 February 2019). Although the sector-specific ministries, such as the Ministry of Health, Nutrition, and Indigenous Medicine, take a leadership role in tackling sector-specific adaptation challenges, the government follows a multi-sectoral approach in developing its strategies (Ministry of Health, personal communication, 13 February 2019). The whole adaptation process was started with all key stakeholders and they identified relevant sectors that were then re-developed by the sector-specific experts and integrated into a comprehensive plan (Ministry of Health, personal communication, 13 February 2019). In general, the multi-sectoral approach works well so far, but it needs to be further strengthened and silos need to be broken up as the outcomes still largely depend on the initiatives of individual actors rather than joint working groups (Ministry of Health, personal communication, 13 February 2019). For climate change in general, eight key sectors exist, and each sector has specific adaptation plans (Ministry of Health, personal communication, 13 February 2019). Additionally, two vulnerability assessments for the health sector will be published in 2019 (Ministry of Health, personal communication, 13 February 2019). Although governmental agencies exist in Sri Lanka as well, policy development seems to take place more within the ministries themselves, supported by renowned scholars in the field; whereas in other countries, such as the UK, governmental agencies take on an even more prominent role in the field of climate change and health.

Changes over time

Already with its first NC, Sri Lanka had a comparatively high recognition score (5 of 7 points), which shows that the country recognized a broad range of climate change related health risks at a time when research on the topic was much less developed than it is now. Moreover, in its first NC in 2000, Sri Lanka already had a significant number of recommendations and other groundwork-level initiatives, which were not part of the first National Adaptation Strategy in 2010 but were included in all subsequent documents.

The number of adaptation-level initiatives in Sri Lanka's health vulnerability assessment from 2010 was extremely high, especially since no concrete adaptation-level measures were reported in any documents before and after. This raises doubts whether the reported adaptation initiatives were actually implemented on the ground. The expert interviews made clear that the adaptation-level score might not be a good representation of the actual implemented health adaptation initiatives in Sri Lanka and needs to be further investigated in future research projects. Sri Lanka's groundwork-level initiatives, however, are relatively stable and have slightly increased over time. What is most striking is the country's early and high awareness of climate change related health risks (Figure 12.4).

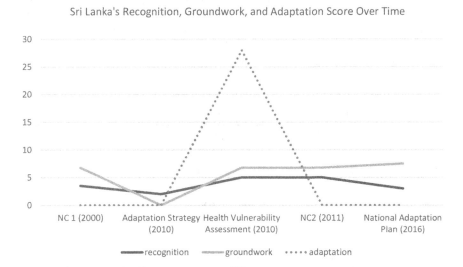

Figure 12.4 Time comparison of health adaptation measures in Sri Lanka

Drivers and barriers

Climate change and health in Sri Lanka – risk assessment

With rank 100 on the ND-GAIN index, an overall vulnerability score of 0.470, and a readiness score of 0.396, ND-GAIN characterizes Sri Lanka as the sixty-seventh most vulnerable country and the ninety-first least ready country (ND-GAIN 2019). The health-specific indicators show very high vulnerabilities in terms of the projected change of cereal yields (ND-GAIN score of 0.963) and the projected change of flood hazard (ND-GAIN score of 0.782). Moreover, with a projected change in vector-borne diseases of 0.658, Sri Lanka's score in this category is much higher than for other countries and shows that recent trends in this category, such as the dengue outbreaks in the late 2010s, will continue to pose severe health risks to the population. Heat-related health risks are less prevalent than in other countries, but still at a comparatively high level with an ND-GAIN score of 0.353. The projected change of deaths from climate change induced diseases (ND-GAIN score of 0.303) and the projected change of sea level rise impacts (ND-GAIN score of 0.259) are at medium levels.

The World Bank's Climate Change Knowledge Portal provides deeper insights into Sri Lanka's exposure to heat-related risks and shows that the projected change in annual probability of heatwave for 2040–2059 will rise to 0.14 under a medium low emission scenario (RCP 4.5) and to 0.21 under a high emission scenario (RCP 8.5) for the same time period. Accordingly, Sri Lanka is more likely to be exposed to heatwaves than other countries, but still does not belong to the most vulnerable countries in this category, such as Indonesia, Venezuela, or Western African countries. Under a high emission scenario, the annual probability of heatwave rises even higher after 2059 to 0.43 between 2060 and 2079 and 0.69 between 2080 and 2099. The annual probability of cold waves will decrease to −0.03 between 2040 and 2059 both under a medium low (RCP 4.5) and a high emission scenario (RCP 8.5). Severe droughts will become slightly less likely in Sri Lanka for the time between 2040 and 2059. Under a medium low emission scenario

Table 12.1 ND-GAIN scores for climate change related health risks in Sri Lanka

Risk	ND-GAIN score
Projected change of cereal yields	0.963
Projected change of deaths from climate change induced diseases	0.303
Projected change in vector-borne diseases	0.658
Projected change of warm periods	0.353
Projected change of flood hazard	0.782
Projected change of sea level rise impacts	0.259

(RCP 4.5), it is projected to decrease to −0.04 for 2040–2059, and 0.00 under a high emission scenario (RCP 8.5).

The risk assessment has shown that the major climate change related health risks for Sri Lanka are related to floods and storms, infectious diseases, and risks related to food and agriculture. Therefore, based on the adaptation portfolio, Sri Lanka focuses on the risks it is actually affected by and there is no significant gap between actual and perceived vulnerabilities and priorities.

Risks assessment by experts from Sri Lanka

The consequences of climate change can be already strongly felt in Sri Lanka (Ministry of Finance, personal communication, 13 February 2019). According to the Ministry of Health, the major climate change related health risks Sri Lanka is facing are vector-borne diseases, especially dengue, extreme weather events, such as floods and droughts, heat-related health risks, and risks associated with air pollution (Dissanayake, personal communication, 13 February 2019; Ministry of Health, personal communication, 13 February 2019). Despite major progress in combatting malaria in Sri Lanka, the risk of dengue has largely increased since around the year 2000 (Ministry of Health, personal communication, 13 February 2019). An exceptionally high number of dengue patients was recorded in 2017 (Ministry of Health, personal communication, 13 February 2019). The number of dengue infections and attributed deaths was close to 200,000 cases of dengue and a little over 400 deaths (Journalist, personal communication, 12 February 2019).

Extreme weather events and shifting rainfall patterns pose severe health risks to Sri Lanka's population through direct and indirect effects (Athukorala, personal communication, 12 February 2019; Ministry of Finance, personal communication, 13 February 2019). In addition to floods and storms, droughts occur more often than in the past, and the combination of floods and droughts puts agriculture, a major pillar of Sri Lanka's economy, under severe stress, which ultimately affects the health of people as well (Ministry of Finance, personal communication, 13 February 2019). Moreover, invasive species, such as the Fall Army Worm, have spread across the country and affect crop yields (Athukorala, personal communication, 12 February 2019). As a consequence of such insects, the government has started to spray pesticides, which again puts people under health risks since pesticides often get into the food chain and lead to long-term health effects (Athukorala, personal communication, 12 February 2019).

Extreme heat, especially due to the urban heat island effect, is among the major climate change related health risks that can be felt in Sri Lanka already and will become even worse in the future (Athukorala, personal communication, 12 February 2019). Construction workers, farmers, and others who spend a lot of time in direct sunlight are particularly prone to such risks (Ministry of Health, personal communication, 13 February 2019).

Risk perception and the political agenda

Overall, awareness for climate change, both amongst the general public and decision makers, has largely increased in Sri Lanka over the past years (Journalist, personal communication, 12 February 2019). Although climate change and related topics are discussed frequently in Sri Lanka, the nexus between climate change and health is not a priority for the government at the moment (Athukorala, personal communication, 12 February 2019). This is partly due to the fact that other topics, such as the recent governmental crisis, dominate the political agenda in general, which leaves less attention for complex topics that climate change and health represent (Amerasinghe, personal communication, 12 February 2018; Athukorala, personal communication, 12 February 2019). The governmental crisis has also led to significant challenges for the bureaucratic system because of the many changes with regard to the political leaders and general guidelines for the respective working units (Journalist, personal communication, 13 February 2019). As a consequence, one interviewed expert claimed that governmental officials were often not able to do their day-to-day work (Journalist, personal communication, 13 February 2019).

Additionally, other consequences of climate change currently receive more attention, such as the effects of changing rainfall patterns on agriculture (Open University, personal communication, 22 February 2019). Although such risks are ultimately also linked to health, this nexus is often not perceived and discussed in the major discourse on the topic (Open University, personal communication, 22 February 2019). Moreover, environmental issues are not always at the forefront of Sri Lanka's political debate, as more short-term oriented issues tend to dominate the agenda (Ministry of Health, personal communication, 13 February 2019). Nevertheless, the topic is frequently discussed in the media and the government takes climate change seriously, as can be demonstrated by its efforts to strengthen its national adaptation framework (Ministry of Health, personal communication, 13 February 2019).

Dealing with complex decisions under uncertainty

Despite Sri Lanka's comparatively high amount of health adaptation initiatives, many experts argue that the government needs to do more to foster cross-sectoral and interdisciplinary cooperation and work less in silos to understand the complexity of climate change related health risks and better adapt to them (Athukorala, personal communication, 12 February 2019). An influential science journalist claimed that one of the major problems is that climate change is often seen as a purely environmental problem, instead of a cross-cutting challenge (Journalist, personal communication, 12 February 2019). Therefore, everything related to climate is passed on to the Ministry of the Environment, which sometimes forwards tasks to the

Ministry of Disaster Management, but agriculture, health, and other sectors are less involved and often work partly isolated from other sectors (Journalist, personal communication, 12 February 2019). Academia is involved in many ways. Sometimes scholars act as consultants for projects and partner with organizations from abroad. However, very often researchers work in isolation and continue with their own research rather than fostering real interdisciplinary cooperation (Open University, personal communication, 22 February 2019). Moreover, despite some comprehensive strategies on adaptation to climate change, much of the government's activities are needs-based and developed very quickly after crises erupt (Amerasignhe, personal communication, 12 February 2019). Much of the long-term policy planning, however, exists more on paper than on the ground, as several experts noted (Amerasinghe, personal communication, 12 February 2019; Journalist, personal communication, 13 February 2019). As a consequence, Sri Lanka has to improve its ability to deal with complex decisions under uncertainty in order to maintain and enhance its health adaptation measures in the future and become more independent from the international community.

The social and cultural dimension

Although Sri Lanka's population is generally aware of climate change, especially since more extreme weather events have hit the South Asian country in recent years, many people do not see the specific risks climate change entails because it is often regarded as a purely environmental issue (Amerasinghe, personal communication, 12 February 2019; Ministry of Health, personal communication, 13 February 2019). This is particularly the case when it comes to climate change related health risks that are often rather abstract and indirect (Amerasinghe, personal communication, 12 February 2019). Often, people see that disease patterns have been changing, but do not know whether such developments are related to climate change (Amerasinghe, personal communication, 12 February 2019; Journalist, personal communication, 12 February 2019). They see the symptoms, but not necessarily the causes of the symptoms and the link between climate change and their everyday life (Athukorala, personal communication, 12 February 2019; Journalist, personal communication, 12 February 2019).

Society's pressure on decision makers to increase their health adaptation efforts varies over time, but overall NGOs tend to cooperate with the government and contribute to concrete proposals rather than protesting and taking to the streets (Journalist, personal communication, 12 February 2019). The Water and Sanitation (WATSAN) Committee, which is one of the few multi-sectoral environmental forums in Sri Lanka, is an example of civil society participation, but even there civil society does not put significant pressure on the government in terms of health adaptation to climate change but rather tries to cooperate and support the government (Athukorala, personal communication, 12 February 2019). From time to time, however, civil society organizations raise awareness for climate change and call for strengthened adaptation action, as was the case after the dengue outbreak in 2017 (Journalist, personal communication, 12 February 2019).

The economic dimension

Overall, the economic development of the country is always at the top of the agenda and issues like climate change are less important to the government and population (Athukorala, personal communication, 12 February 2019). Within the climate change debate, economic topics, such as the impact of climate change on agriculture, are at the forefront of the debate, whereas health does not play a prominent role (Open University, personal communication, 22 February 2019). Due to the fact that Sri Lanka is a developing country, more immediate policy areas often take precedence when it comes to planning the country's budget, which is why the country often depends on the international community to provide the funds for health adaptation work (Ministry of Finance, personal communication, 13 February 2019).

The international dimension

The international community plays a very important role in Sri Lanka's health adaptation initiatives since international organizations, such as UN Environment, UNDP, and the WHO have been active in the country for a long time and are involved in a number of projects that promote adaptation to climate change (Amerasinghe, personal communication, 12 February 2019; Dissanayake, personal communication, 13 February 2019; Journalist, personal communication, 12 February 2019; Ministry of Health, personal communication, 13 February 2019). They provide technical and financial assistance and thus have a significant influence on the processes and outcomes of the specific projects (Ministry of Health, personal communication, 13 February 2019). Additionally, Sri Lanka receives funding from international climate funds, such as the Green Climate Fund and funds provided by the World Bank (Ministry of Health, personal communication, 13 February 2019).

Although the World Bank is not currently involved in a project that would specifically addresses climate change and health, one project that is currently being prepared is tailored towards improving impact-based forecasting tools and early warning to strengthen the meteorological department's forecasting abilities, which will affect the health sector as well (Dissanayake, personal communication, 13 February 2019). Additionally, the World Bank has an ongoing project on water supply and sanitation and a Health System Strengthening Project in Sri Lanka that targets non-communicable diseases (Dissanayake, personal communication, 13 February 2019). Moreover, in recent years, many flood mitigation programs have been supported by the

World Bank (Ministry of Finance, personal communication, 13 February 2019).

Furthermore, international organizations support the government in preparing its NCs to the UNFCCC, their Nationally Determined Contributions, and other key documents that inform about Sri Lanka's progress on climate action (Ministry of Health, personal communication, 13 February 2019). Additionally, the international community helps to put climate change on the domestic political agenda (Ministry of Health, personal communication, 13 February 2019). According to an expert from the Ministry of Health, international conferences and treaties, such as the Paris Agreement, send signals to the government and the population to take comprehensive action against climate change (Ministry of Health, personal communication, 13 February 2019). At the same time, however, the expert claimed that health has not been a primary topic at COPs and other international climate change conferences yet, which shows that the topic has much potential to develop further if it receives a more prominent place on the agenda of COPs and other conferences (Ministry of Health, personal communication, 13 February 2019).

Although it has become clear that the international community has significantly supported the development of health adaptation measures in Sri Lanka, two major areas of criticism exist. First, while international experts often cooperate with local consultants and agencies, there seems to be a clear gap between the strategies on paper and the actual actions on the ground (Athukorala, personal communication, 12 February 2019; Journalist, personal communication, 13 February 2019). Second, many measures are project based and depend on international funding, which means that, after the projects end, the governmental agencies often do not continue with the measures (Open University, personal communication, 22 February 2019). Accordingly, due to the high dependence of the projects on international funding, they cannot be characterized as sustainable. To ensure real sustainability and effectiveness, there needs to be a transition from project ownership to policy ownership on the side of the Sri Lankan government.

Summary

Sri Lanka's high CHAIn score represents what the country claims it is doing in terms of health adaptation to climate change, but there is a significant gap between its groundwork-level and adaptation-level initiatives. Whilst Sri Lanka's groundwork-level initiatives are rather robust and stable over time with a steady increase over the years, its adaptation-level initiatives were identified basically in one document and many experts claim that Sri Lanka's implemented adaptation actions are much lower in reality. Nevertheless, Sri Lanka's early and high awareness for various climate change related health risks and its high number of groundwork-level initiatives represent a remarkable level of engagement in comparison with countries of a similar economic level. Moreover, Sri Lanka mostly addresses the specific risks it is actually affected by, which include infectious diseases, risks related to floods and storms, and risks related to food and agriculture.

Sri Lanka's high score on the CHAIn index is particularly surprising as the governmental institutions and the research capacity of the country are not more developed than in other countries and, despite some multi-sectoral work in the field, no significant institutional drivers exist that would foster health adaptation to climate change. Although some actors within the ministries put a lot of effort into Sri Lanka's health adaptation work, its capacities are much more limited compared with those of other countries. From an economic perspective, Sri Lanka does not have the capacity at the moment to invest significantly more in addressing complex policy challenges because more urgent issues, such as developing infrastructure and fostering agriculture and tourism, are of higher relevance to the country. As a consequence, climate change and its implications for health are not frequently at the top of the agenda, especially against the background of the recent governmental crisis. Additionally, the pressure from the Sri Lankan society to step up adaptation efforts varies over time and mostly increases when extreme weather events or rises in vector-borne diseases occur. Overall, the public understanding of the nexus between climate change and health is rather limited, which is why it does not put significant pressure on the government in this field.

This goes along with the finding that the international community has a very strong impact on health adaptation to climate change in Sri Lanka and partially takes over functions that the state is supposed to perform. International organizations help to develop strategies and plans. UN organizations, such as the WHO, and international funds, such as the World Bank, work on concrete projects with the country to strengthen its adaptation work. Despite a lot of progress in the field of health adaptation to climate change thanks to international cooperation, the critical questions remain as to how sustainable such cooperation is and what happens after the funding for individual projects runs out. The key to success seems to be developing joint and equitable projects from the beginning that are designed to develop sustainable solutions and ensure policy ownership even after the project ownership terminates.

References

ND-GAIN. 2019. "Sri Lanka." Accessed 23 May 2019. https://gain-new.crc.nd.edu/country/sri-lanka.

13 General findings

The case studies of the UK, Ireland, the RoK, Japan, and Sri Lanka have shown how differently states perceive and respond to climate change related health risks. The insights have led to numerous general findings on the drivers of and barriers to health adaptation to climate change, which will in the following be discussed and subsequently be utilized to update and strengthen the theoretical concept of health adaptation to climate change of this thesis.

The UK, the highest-ranking country on the CHAIn, has largely expanded its work on health adaptation to climate change after the adoption of the Climate Change Act in 2008, which can be traced back to the lessons the country has learned from extreme weather events in the early 2000s, especially the massive heatwave in 2003. Besides some important facilitating factors for the institutional development of climate change adaptation in the UK, such as high societal awareness and pressure from civil society in the early 2000s, as well as the rather strong economic situation at that time, the major driver of health adaptation to climate change in the UK is the epistemic community in the country, especially with regards to the strong researchers, think tanks, and advocacy coalitions around the Lancet Countdown on Climate Change. Since the UK is home to some of the world's best universities and research institutes, the research community influences the societal and political perception of climate change, especially when it comes to attributing extreme weather or temperature events to climate change. Although the economic situation has deteriorated and the political agenda has drifted away from climate change since 2008, especially with Brexit talks, the institutions that were created with the Climate Change Act have helped the UK to strengthen and expand its health adaptation work. Moreover, the international community, especially through IPCC publications or pressure from the EU, helps to put the topic on the agenda from time to time and re-evaluate the government's work in the field.

The situation in Ireland has been quite different. The country has only recently started to expand its adaptation work to climate change and will update its sectoral strategies in 2019. Although a small epistemic community in the field of climate change and health exists, it has not been able yet to

Table 13.1 Climate change and health risks, risk perception, policy responses, drivers and barriers across case studies

	UK	Ireland	South Korea	Japan	Sri Lanka
Vulnerability	Low	Low	Low	Low	High
CHAIn score	109	9	103.25	13.75	46.25
Recognition	7	6.5	5.5	4.5	5.5
Groundwork	42	0.5	9.75	7.25	12.75
Adaptation	60	2	88	2	28
Drivers	Attribution of extreme weather events, epistemic community, advocacy groups, international community	Attribution of extreme weather events, international organizations, esp. the EU	Attribution of extreme weather events, strong governmental agencies (KACCC), epistemic communities, international organizations, export-oriented economy	Attribution of extreme weather events, international organizations	Attribution of extreme weather events, epistemic community, international organizations
Barriers	Brexit, economic crises	Focus on other topics (e.g. health system crisis), lacking capacity, economic crises, complexity and uncertainty	–	Focus on other topics, negative image of adaptation compared to mitigation, lacking policy integration, complexity and uncertainty	Economic capacity, complexity and uncertainty
Milestones	Climate Change Act (2008)	Climate Action and Low Carbon Development Act (2015)	Establishment of enter for Climate Change Adaptation (KACCC) (2009)	Climate Change Adaptation Act (2018)	National Climate Change Act (announced in 2019)

significantly frame the public discourse on climate change and health and civil society's pressure on the government is rather low. With the Climate Action and Low Carbon Development Act from 2015 and subsequent adaptation process, the government has started to expand its adaptation work, but in international comparison it is still rather late when it comes to national-level health adaptation to climate change. The international community, especially the EU, frequently set the topic on the agenda and put pressure on the government to evaluate and enhance its adaptation work. In the end, however, the government's willingness to adapt largely depends on how it perceives the economic situation of the country and whether it believes that health adaptation to climate change is a meaningful investment in relation to other policy areas.

As a global leader in adaptation-level initiatives to climate change related health risks, the RoK has demonstrated how the institutionalization of multi-sectoral adaptation work through governmental agencies, such as the KACCC, can lead to improvements concerning the coordination and development of health adaptation measures. The government's risk perception and lessons learned from the extreme weather events in the early 2000s, largely influenced by the country's epistemic community and pursuit of international reputation, have led to the establishment of new institutions that strengthened the country's adaptation work. Although civil society did not put more pressure on the government with regards to climate change adaptation than in other countries, the RoK has developed a considerable number of adaptation-level initiatives. This is mostly due to the strong indirect influence of the international community and the RoK's pursuit of international reputation. Moreover, the attribution of extreme weather events to climate change has led to learning experiences and new institutions to coordinate the government's work.

Japan, on the contrary, has only recently started to strengthen its health adaptation measures, which is at least partly due to the negative public image climate change adaptation had in the past, compared with mitigation. Since parts of the society feared that a greater emphasis on adaptation might hinder mitigation efforts, societal pressure on the government in this field was relatively low, despite the fact that extreme weather events occurred frequently and were attributed to climate change. Although some experts on climate change and health are active in Japan, the country does not have a strong epistemic community that frames the discourse and contributes to strengthened health adaptation measures and advocacy coalitions between NGOs and researchers are less strong than, for instance, in the UK. The international community's influence is rather indirect and waxes and wanes in relation to international conferences, such as COPs, and IPCC publications. As extreme weather events have further increased in recent years, especially in 2018, the government has started to expand its adaptation work and it can be expected that the country's CHAIn ranking will increase.

Compared with other countries, Sri Lanka ranks exceptionally high on the CHAIn. However, the case studies have shown that the health adaptation measures the country reported in its national-level strategies and plans are not always followed up with concrete measures on the ground. Moreover, the country's high ranking can largely be explained by the influence of international organizations, because the World Bank, the UNDP, the WHO, and others finance and implement concrete adaptation projects and support the government in developing strategies and plans with external consultants. The epistemic community is less strong in Sri Lanka than in other countries, but it cooperates with the international community and contributes to the adaptation planning process.

Overall, the case studies have shown that the risk perception and attribution of health risks to climate change have a significant impact on states' health adaptation work. Due to the high complexity and uncertainty of the topic, the perception and adaptation-willingness are largely influenced by epistemic communities and international organizations. In addition to framing the topic and contributing to setting the agenda, in developing countries, such as Sri Lanka, international organizations directly influence health adaptation to climate change by providing funding, strategy development support, and co-creating concrete adaptation-level projects. In developed countries, their influence is rather indirect and often increases with international events and publications on the topic. Countries that pursue international reputation, such as the RoK, are more receptive to developments in the international climate change regime than others, but overall, IPCC publications especially are broadly received and cited across countries.

Part IV
Conclusion

14 Main findings

How do states perceive and adapt to climate change related health risks? Which factors influence states' perception of and measures to adapt to climate change related health risks? These were the two research questions that drove and structured this thesis. To answer the first question, the CHAIn and its sub-indexes – the recognition-level, groundwork-level, and adaptation-level indexes – were developed to track health adaptation measures of 192 states and compare the levels and types of adaptation as well as which risks states address in their national-level documents on adaptation to climate change.

The results show great differences in terms of which climate change related health risks states recognize and which actions they take to address them. The breadth and depth of adaptation measures vary to a great extent, which is reflected even among the top ten states on the CHAIn: with a total score of 109 points, the UK ranks first, followed by the RoK with 103.25 points. While countries such as Japan (rank 85), Sweden (122), and Ireland (123) rank in the middle of the index, the last ten states include many countries that, based on other indexes, such as GDP (PPP) or GDP (PPP) per capita, were expected to rank higher. A closer look at the individual sub-indexes reveals that, except for North Macedonia, all high-ranking states have already implemented adaptation-level measures to protect their populations from climate change related health risks, such as infrastructure measures or awareness campaigns. With an adaptation score of 88, the RoK claims in its national documents to have implemented an extraordinarily high number of adaptation-level measures to which only the UK (adaptation score of 60) and Cyprus (adaptation score of 48) can even compare. The groundwork scores of countries also vary to a great extent. Although the RoK ranks second on the overall CHAIn, it has a groundwork score of only 9.75, which is significantly lower than most of the other top ten countries on the CHAIn. The recognition-level index demonstrates that only a few countries, such as the UK, Jordan, and Canada, recognize all climate change related health risks that are currently discussed in academia.

Overall, the vast majority of health adaptation measures (both groundwork-level and adaptation-level) address the general nexus between climate change and health. The second-most addressed risk category is "infectious diseases," with 736 groundwork-level initiatives and 149 adaptation-level initiatives. The distribution shows that states' understanding of the nexus between climate change and health is still rather narrow. Tertiary climate change related health risks, such as risks related to migration, conflicts, and mental health, are addressed by only a very few states. In terms of adaptation types, the vast majority of health adaptation measures constitute recommendations (2,514 out of 4,271 reported initiatives), strategies and plans (659 initiatives), and measures in the category "research and development" (488 initiatives). Adaptation-level initiatives are largely underrepresented compared with groundwork-level initiatives. This shows that the international community is still at the beginning of grasping the challenge and adapting to climate change related health risks.

The Sieve Model of health adaptation to climate change

The CHAIn and its sub-indexes have revealed some striking differences in terms of which climate change related health risks states recognize and what they do to address them. The tracking of health adaptation measures in this thesis helps states and international organizations to learn from each other and identify strengths and weaknesses. But how can the differences among states be explained? Why have some states already developed rather extensive strategies and plans and even implemented adaptation-level initiatives, while others do not even acknowledge the nexus between climate change and health in their national-level climate change documents?

Climate change is a very complex topic with high levels of uncertainty. Therefore, adaptation to climate change is often called a "wicked problem." The nexus between climate change and health can be considered even more complex and uncertain since climate change is only one of many factors that affects health and often exacerbates various health determinants that are hard to identify (Wilkinson, Campbell-Lendrum, and Bartlett 2003, McMichael 2014). Moreover, research on climate change and health has only recently garnered more significant attention from the scientific community, and many unknowns still exist. Due to the high complexity and uncertainty associated with the nexus between climate change and health, as well as the rapid nature of politics, which often focuses on more immediate crises than the climate crisis, states depend on epistemic communities to help gather information on this long-term policy challenge, understand the risks, and develop appropriate measures to reduce them. The information that states receive on climate change and health, how they perceive the information, and how they use it to prepare for the future are all dependent on a number of significant factors.

First and foremost, the more political rights and civil liberties states guarantee – in other words, the more democratic states are – the more climate change related health risks they recognize. This is mostly due to the greater competition of different opinions between civil society groups, interest groups, and, importantly, epistemic communities. Additionally, the more democratic states are, the more likely they will be to push politicians to acknowledge that climate change exists and that it affects the health of populations. At the same time, both democratic and autocratic states are rather cautious with regards to the level of attribution since they do not intend to create any specific policy implications by directly attributing certain diseases or mortality rates to climate change. In states with stronger interest groups that advocate for health adaptation to climate change, such as through civil society organizations or specific epistemic communities, states may increase their attribution and recognition levels, which may ultimately lead to higher amounts of adaptation as well. Moreover, interest groups within states often frame how extreme weather events, such as floods, storms, and heatwaves, are perceived by states and whether they are attributed to climate change. States' response to climate change related health risks depends less on what experts in the field may describe as vulnerability or risk exposure, but more on how they perceive climate change related health risks and what consequences this has for their actions.

Concurrently, international organizations, such as UN Environment, UNDP, or the WHO, may spread information on climate change and health, influence interest groups within states, and further influence states' recognition and perception of climate change related health risks, which may lead to higher recognition scores and the topic rising to the top of the political agenda. Moreover, particularly in developing countries, where international organizations provide significant funding, they may directly influence how topics are discussed and priorities are set, especially in relation to such complex policy challenges as climate change and health since international organizations often bring expertise that states are lacking.

If states are aware of climate change related health risks and willing to take measures to act against them, they need the appropriate resources to do so. Therefore, it can be stated that the higher the GDP of a state is, the more adaptation-level measures it will implement. Adaptation resources, however, constitute not only financial measures to invest in health adaptation, but also expertise and experience in the field. Due to the high complexity and uncertainty of the topic, epistemic communities often provide the expertise that states need to develop health adaptation measures. The case studies have shown that in states that do not have such epistemic communities or adequate capacities within governmental institutions, such as governmental adaptation centers like the KACCC in the RoK, international organizations often provide strategy and planning services or work on concrete adaptation-level projects with governments. Moreover, as was the case in Cyprus and the Republic of Moldova, other states such as the UK or Germany may provide funding and policy support to these states to develop strategies, plans, and concrete measures to adapt to climate change related health risks. For states that possess the resources required to adapt, the international community plays a more indirect role by helping to put the topic on the agenda through international conferences, such as the COPs, or publishing reports on the topic, such as the IPCC assessment reports.

Overall, the major drivers of health adaptation to climate change are the strength and influence of epistemic communities and international organizations in spreading information on climate change and health, framing perceptions of climate change related health risks, and developing health adaptation measures with or for states. If international organizations take on state functions in cases where states are not able to develop appropriate measures, the sustainability of such projects is often in question. The key to success is to ensure that states receive policy ownership through the projects they conduct with international organizations and to design such projects to stay alive after funding and support from the international community can no longer be provided.

Refining and implementing the model

As one of the first large-scale political science studies on climate change and health, this study envisioned achieving the following:

a   To offer a starting ground for interdisciplinary research on climate change and health that incorporates perspectives from political science and public health literature
b   To establish a global index that compares the health adaptation initiatives of all UN Member States
c   To develop an innovative mixed methods design that connects quantitative and qualitative research traditions to contribute to a better understanding of the drivers of and barriers to health adaptation to climate change from a political science perspective

Through its interdisciplinary approach and the numerous cooperation projects and exchanges with scholars from public health research, physics, meteorology, and many other disciplines, this study has shown that proper interdisciplinary cooperation is possible, beneficial, and absolutely necessary for cross-cutting and complex topics such as the nexus between climate change and health. The thesis seeks to inspire further political science-based research on the topic and spark interdisciplinary cooperation in the field to create win–win situations for academia, society, and politics.

The entire book, but especially the CHAIn and its sub-indexes, has started to fill a wide research gap in the field of health adaptation to climate change and contributed to a better understanding of how states perceive climate change related health risks and which measures they take to protect their populations. In addition, states, international organizations, the private sector, civil society, and essentially every citizen can learn from this study about how different states adapt to climate change related health risks, which best practices should be widely implemented, and where further efforts are urgently needed.

Founded on a pragmatic approach to political science research, the mixed methods design explores new horizons in terms of the combination of various ontological, epistemological, and methodological traditions to find answers to new and complex research challenges by benefiting from the individual strengths of different approaches and methods whilst at the same time balancing out their respective weaknesses. While the study has led to a better understanding of how and why states adapt to climate change related health risks, the thesis lays the foundation for a great number of further research projects in the field.

First and foremost, further research should enhance the CHAIn and its sub-indexes by expanding the database; adding time-series comparisons; and updating, testing, and re-evaluating the categories and their weights. In a constantly changing and rapidly developing research area, new climate change related health risks are discovered on a very frequent basis, and states' perceptions and policies develop rapidly. For the first time, this study has tracked initiatives for health adaptation to climate change of 192 states. It is now up to further research projects to push this groundwork to the next level. Furthermore, despite significant findings, the explanatory power of the regressions can be improved by theorizing and testing further potential explanatory variables that may influence how and why states adapt to climate change related health risks. Additionally, different proxies to operationalize the impact of epistemic communities and international organizations on national-level health adaptation measures should be considered to further strengthen the explanatory model.

Climate change does not know any boundaries. To understand the challenge and develop solutions, all actors need to cooperate and overcome their own mental barriers. Realizing that barriers are what we make of them is an important starting point. To overcome these barriers, we need innovative and constructive forms of cooperation without boundaries.

References

McMichael, Anthony. 2014. "Climate change and global health." In *Climate Change and Global Health*, edited by Colin Butler, 11–20. Canberra.

Wilkinson, P., D. H. Campbell-Lendrum, and C. L. Bartlett. 2003. "Monitoring the health effects of climate change." In *Climate Change and Human Health – Risks and Responses*, edited by A. J. McMichael, D. H. Campbell-Lendrum, C. F. Corvalán, K. L. Ebi, A. K. Githeko, J. D. Scheraga and A. Woodward, 204–219. Geneva: World Health Organization.

15 From the climate crisis to the health crisis and vice versa – a COVID-19 update

The COVID-19 pandemic has changed the world as we knew it. Within days, the SARS-CoV-2 virus, which was first recognized in Wuhan, China, in December 2019, spread across the world, ultimately leading to millions of infections (WHO 2020a). At the time of writing this chapter, over 41 million people are infected with COVID-19 and over 1.1 million people have died from the disease (WHO 2020b). Concurrently, the pandemic has drastically affected the socioeconomic situation in almost all countries in the world (World Economic Forum 2020c). Inequalities have risen and governmental priorities have shifted even more to responding to immediate crises rather than long-term policy challenges (World Economic Forum 2020b). Some countries, however, such as Germany and the Republic of Korea, combined climate change mitigation measures with their crisis response by initiating sustainable recovery programs (World Economic Forum 2020a).

Looking back at the first half of 2020, it is safe to say that the COVID-19 pandemic has led to numerous challenges and, at the same time, created new opportunities. In this context, at least one aspect has become clear: where there is willingness to act, even measures that were previously described as unimaginable become possible. Accordingly, with regards to the five case studies and the overall findings of this book, a number of new questions emerge in light of the recent pandemic:

a   Have states associated the COVID-19 pandemic with climate change?
b   What role have climate change related health risks played in governmental responses to the COVID-19 pandemic?
c   Have governments strengthened their health adaptation measures to climate change as a consequence of the COVID-19 pandemic?
d   Have governments emphasized trade-offs or synergies between economic recovery measures and climate change mitigation?
e   What role does the international community play with regard to national-level measures on COVID-19 and climate change?
f   What can we learn from governmental responses to COVID-19 for decision making on climate change related health risks?

To answer these questions, I conducted a detailed, structured analysis of all governmental documents and articles in key national newspapers in the five case studies issued between February and August 2020, using the key-words "climate change" *and* "COVID-19." Overall, 892 governmental documents, such as speeches by ministers and heads of governments, press statements, and policies, were analyzed. Additionally, 1,738 newspaper articles were part of the data set.[1] In general, the analysis shows that while some states have drawn the link between climate change and COVID-19, none of them has announced direct consequences in terms of health adaptation to climate change. Climate change mitigation, on the other hand, was prominently discussed and integrated into recovery measures in some states, especially the Republic of Korea.

The United Kingdom of Great Britain and Northern Ireland

The majority of the UK's governmental statements on the nexus between COVID-19 and the climate crisis was issued by representatives of the upcoming COP26 Presidency and the Environment Agency. In their reports, press releases, and speeches, governmental officials acknowledged the link between climate change and the pandemic (UK Department for Business, Energy & Industrial Strategy 2020a, b, UK Environment Agency 2020). Moreover, they frequently emphasized the business opportunities of a green recovery and synergies between climate change mitigation, especially through sustainable investments, and COVID-19 recovery programs, as the following quote from COP26 President Alok Sharma illustrates:

> Of course, countries around the world have responded with urgency to helping their populations through the Covid pandemic, as we have done in the UK. That is absolutely the right approach. And as we turn our attention to the economic fightback, we need to also reflect this desire to address the climate crisis in our economic response.
> (UK Department for Business, Energy & Industrial Strategy 2020a)

Similarly, at a webinar of the London Stock Exchange, Alok Sharma emphasized: "So as we emerge from the public health pandemic, we not only want to support businesses to bounce back as quickly as possible, but also to do so in a way that meets the UK's big, structural challenges" (UK Department for Business, Energy & Industrial Strategy 2020b). As the official statements and the overall line of communication demonstrate, synergies between climate change mitigation and measures to restart the economy have often been addressed by government representatives, and plans for a green recovery are also part of the UK's overall new deal to "build, build, build" the economy after COVID-19 (Prime Minister, UK 2020). Concrete actions, such as the amount of investment in climate change mitigation or adaptation measures, however, have not been further specified by the government.

While the British government has directly linked the health crisis due to COVID-19 with climate change mitigation, climate change related health risks and the need for strengthened health adaptation measures have not been discussed in any of the official documents since the outbreak of the virus in early 2020. Although many of the statements by the British government on climate change and COVID-19 were made in the context of the upcoming COP of the UNFCCC, the international community, such as UN organizations or the European Union, were not explicitly mentioned in the official statements and did not seem to significantly influence the government's policies. To put it in a nutshell, the UK definitely connected COVID-19 with climate change, especially when it comes to business opportunities of a green recovery. Concrete actions, however, such as sustainable investments or capacity building measures, are yet to be discussed and adopted.

The Republic of Ireland

In the past few months, the strongest proponent for strengthening climate action and learning from the COVID-19 pandemic in Ireland has not been a current, but a former, member of the government, namely Mary Robinson, who was Ireland's President from 1990 to 1997 (O'Sullivan 2020). During her address in March 2020, she argued that we should use the direct and effective response towards COVID-19 to also strengthen climate action (O'Sullivan 2020). The Irish government, on the other hand, did not mention the nexus between climate change and health in any official document at the beginning of the pandemic, nor did it bring up the term green recovery. In June 2020, however, after four months of negotiations, Ireland formed a new government (Guardian 2020). For the first time in history, the two civil war parties, Fianna Fáil and Fine Gael, formed a coalition with the support of the Green Party (McQueen 2020). While the new government agreed on setting the goal of reducing greenhouse gas emissions by 7 percent annually, it did not announce any COVID-19 related changes in its climate policy (*Irish Independent* 2020).

As a consequence, the analysis has shown that Ireland has not made a link between COVID-19 and climate change in its official governmental documents since the outbreak of the disease. It has neither mentioned climate change related health risks in the context of COVID-19 nor strengthened its health adaptation measures as a consequence of the pandemic. Synergies or trade-offs between climate change mitigation and COVID-19 have not been addressed publicly and the international community has not significantly influenced national-level decisions on COVID-19 and climate change.

The Republic of Korea

On 15 April 2020, in the midst of the COVID-19 pandemic, the Republic of Korea (RoK) held legislative elections (Vetter 2020). With 180 out of 300 seats in the National Assembly, the Democratic Party of President Moon Jae-in celebrated a clear victory and gained 60 seats more than in the previous election (Vetter 2020). What made the elections special, however, is the fact that the success of the governing party was largely driven by the government's response to COVID-19 and its plan to implement a Green New Deal (Vetter 2020). Originally outlined in a campaign manifesto, the Green New Deal has become the RoK's plan to move from a fossil-fuel dependent to a low-carbon economy by investing 114 trillion won (around 85 billion €) by 2025 (Korea 2020d, 2020a). In fact, together with local governments and the private sector, the Korean government seeks to increase the investment to 160 trillion won (119 billion €) (Korea 2020a). Through the investment, more than 890,000 new "green jobs" shall be created by 2022 and a total of 1.9 million by 2025 (Korea 2020a).

In their official statements, members of the Korean government repeatedly linked the Green New Deal with COVID-19 and connected their response to the climate crisis with the global health crisis the world has been experiencing in 2020 (Korea 2020a). In a keynote address in July 2020, President Moon Jae-in stated: "The Green New Deal is about responding preemptively to the climate crisis, a desperate reality already confronting us. The COVID-19 pandemic has reaffirmed the urgency of responding to climate change" (Korea 2020a). Furthermore, the government stressed the importance of climate change mitigation for improving health and well-being, especially when it comes to the prevention of infectious diseases like COVID-19: "Amid a consensus that climate change responses are also indispensable for preventing infectious diseases, advanced countries in Europe and elsewhere have already been pushing a Green New Deal as a key task" (Korea 2020a). At the same time, the link between climate change and health risks stayed rather broad in the government's public response to COVID-19 and, except for calls to build up infectious disease response systems, no references have been made to protect the population from climate change related health risks. Overall, the focus of the government's response is much more on climate change mitigation than health adaptation to climate change. With regards to mitigation measures, such as those integrated into the Green New Deal, synergies between climate change mitigation and economic recovery measures have been repeatedly highlighted by the government (Korea 2020d, 2020a, *Korea News Gazette* 2020).

The international community, especially the EU, has played an outstanding role in the formation of the Korean Green New Deal. In his statements, President Moon Jae-in repeatedly referred to the EU's Green Deal as an inspiration and the EU as a whole as an essential partner in the field:

> I would like to convey my respect to the new EU leadership for being at the forefront of addressing global climate and environmental issues through the European Green Deal policy. I hope the EU will become an important partner for my government's Green New Deal policy.
>
> (Korea 2020c)

The analysis has confirmed that the RoK's standing in the international community is of great importance to the government and, accordingly, has been a strong motivating factor behind the creation of the Green New Deal. In an official meeting in June 2020, President Moon Jae-in made clear that "[t]he Government will pave the way toward sustainable growth through the Green New Deal. We will create new markets, industries and jobs while actively responding to climate change as a responsible member of the international community" (Korea 2020b). As with the RoK's health adaptation policies, when it comes to its COVID-19 response, the country takes its standing in the international community very seriously since the government directly connects its international reputation with international trade and general business opportunities (Korea 2020b). The RoK case thus shows that other states and the international community can be a significant source of policy diffusion and motivation for others to act on long-term policy challenges, as described in the Sieve Model.

Japan

"Individual actions and international cooperation are both indispensable to fight against our common enemies, COVID-19 and climate change" (Koizumi 2020b). Speaking at the Petersburg Climate Dialogue in April 2020, Japan's Minister for the Environment, Shinjiro Koizumi, acknowledged the nexus between COVID-19 and climate change. At the same time, however, Japan's actions to tackle both COVID-19 and climate change have been rather limited so far. Although Koizumi repeatedly underlined the importance of overcoming both the climate crisis and the current global health crisis, the first specific action that incorporated climate change in the state's COVID-19 response was announced only in August 2020, much later than in other states across the globe (Ministry of the Environment, Japan 2020b).[2] With the launch of the online platform "Sustainable and Resilient Recovery from COVID-19" and the website "Platform for Redesign 2020" in August 2020, the Japanese government initiated a new form of international collaboration on a green recovery from COVID-19, which integrates discussions and the exchange of knowledge on various areas of climate change mitigation and adaptation (Koizumi 2020b, 2020a, Ministry of the Environment, Japan 2020b). The international community has played an important role for Japan's response on COVID-19 and climate change. The online platform on sustainable recovery was launched at the UNFCCC's Momentum for Climate Change event, and international organizations, especially the various UN bodies, were mentioned several times in the government's statements on the topic (Koizumi 2020a, Ministry of the Environment, Japan 2020c). Health adaptation to climate change, however, was not specifically mentioned in any of the governmental documents between February and September 2020.

Overall, the analysis has shown that climate change related health risks have not been directly addressed in the Japanese government's response to the COVID-19 pandemic and the government has therefore not yet strengthened the health adaptation measures as a consequence of the pandemic. A debate on synergies and trade-offs between climate change mitigation and recovery mechanisms has barely taken place in the official discourse. In some statements, however, Japan's Minister of the Environment emphasized the importance of a green recovery from COVID-19. Nevertheless, except for the establishment of an international knowledge-sharing platform, the government has not taken any specific actions on the subject. As with Japan's general health adaptation policies, in the case of COVID-19, Japan was rather late with making the link between climate change and health in their official governmental documents. However, once other countries and international organizations started pushing for green recovery programs, the Japanese government referred to the importance of the nexus between climate change and COVID-19 in speeches and press releases and initiated an online platform for international exchange and knowledge transfer on the topic. Considering that Japan has not addressed climate change related health risks in official statements following the COVID-19 outbreak, it remains unclear how the pandemic has affected their risk perception in this field.

Sri Lanka

In comparison with the other case studies, the number of statements by Sri Lankan (?) public officials on the nexus between climate change and COVID-19 has been very limited. The only official governmental record, in which climate change was mentioned in the context of COVID-19, was a statement by Foreign Secretary Admiral Professor Colombage to the BIMSTEC Member States (Governmental News, Sri Lanka 2020). Although not directly asking for a green recovery program or health adaptation to climate change, he urged BIMSTEC member states to continue cooperation in the field of trade and development while taking into account the effects of climate change on their respective countries (Governmental News, Sri Lanka 2020). In addition to the official governmental source, an interview by the *Sunday Observer* (Sri Lanka) with Dr. Inoka Surweera, Consultant Community Physician from the Environmental and Occupational Health Directorate of the Ministry of Health, was published in June 2020 that referred to climate change and COVID-19 (*Sunday Observer* 2020). However, Dr. Surweera only briefly mentioned the importance of trees and forests in combatting climate change; he did not refer to climate change related health risks or concrete actions that the government has taken or should take against the backdrop of the COVID-19 pandemic (*Sunday Observer* 2020).

Considering that no other official governmental documents referred to climate change and COVID-19 between February and September 2020, it can be stated that the Sri Lankan government did not officially connect the COVID-19 pandemic with climate change, and climate change related health risks were not addressed in the country's response to the pandemic. Moreover, the government did not strengthen its health adaptation measures as a consequence of the pandemic, and a debate on synergies or trade-offs between economic recovery measures and climate change mitigation did not take place. Furthermore, the international community did not significantly influence Sri Lanka to strengthen their climate response as a consequence of COVID-19.

Lessons learned?

The measures decision makers all over the world have taken to counter the spread of COVID-19 have powerfully demonstrated how far governments are willing to go to protect their populations from health risks. Moreover, the amount of investment to recover from the economic crisis due to COVID-19 has shown that financial resources can be made available very quickly to tackle massive challenges if the political will exists. If we apply these two lessons to climate change related health risks, there is reason to hope for strengthened mitigation and adaptation measures in the near future to protect both planetary and human health. Recent debates and actions for green recovery programs in countries such as the RoK and the UK have shown that, at least in some countries, sustainability is moving from sector-specific modules and targeted initiatives to comprehensive approaches that are mainstreamed across (nearly) all policy sectors. Nonetheless, sustainability is far from being a mainstream approach and priority in all countries across the world and a gap between knowledge, debate, and feasible actions still exists.

The findings on COVID-19 and climate change reinforce the overall findings of this thesis, especially as presented in the Sieve Model of health adaptation to climate change: many of the global challenges humanity is facing in the twenty-first century are complex, wicked problems. Innovation to solve these challenges often originates in small circles of society, most frequently in epistemic communities that are internationally connected, equipped with the knowledge and skills to understand the challenges, and motivated to, at least temporarily, step out of their role as pure investigators into that of advisors and advocates for change. Once different actors within epistemic communities unite and effectively cooperate with international organizations and governments, information on climate change related health risks can move to the next level of diffusion within the political system. When internal and external factors, such as formative events, governmental changes, or developments within the international community, add pressure on decision makers, their willingness to act increases and, if they have the necessary resources to act, policy windows open and political change becomes possible. While this is not yet the case for climate change related health risks, international reactions to the current health crisis have laid bare what is possible if risk perceptions change.

Notes

1 All relevant passages of the analyzed documents can be found in the supplementary material.
2 In July 2020, the Japanese government released its roadmap to "Beyond-Zero" Carbon, in which it outlined Japan's strategy to achieve its greenhouse gas reduction goals through mitigation and carbon capture and storage measures (Ministry of the Environment, Japan 2020a). However, despite briefly mentioning COVID-19, the government did not directly link its measures to the pandemic.

References

Governmental News, Sri Lanka. 2020. "Sri Lanka urges BIMSTEC Member States to embrace the "new normal" for regional prosperity." 4 September 2020. https://www.un.int/srilanka/news/sri-lanka-urges-bimstec-member-states-embrace-%E2%80%9Cnew-normal%E2%80%9D-regional-prosperity.

Guardian, The. 2020. "Ireland to form new government after Green party votes for coalition." *The Guardian*, 26 June 2020. https://www.theguardian.com/world/2020/jun/26/irish-government-to-be-formed-after-greens-vote-yes-to-coalition.

Irish Independent, The. 2020. "Key points in government formation deal." *The Irish Independent*, 15 June 2020. https://advance-lexis-com.ubproxy.ub.uni-heidelberg.de/api/document?collection=news&id=urn:contentItem:604N-7B31-DYTY-C2KM-00000-00&context=1516831.

Koizumi, Shinjiro. 2020a. "June Momentum opening session video message by KOIZUMI Shinjiro, Minister of the Environment, Japan." 1 June 2020. https://www.env.go.jp/en/earth/message_text_June_Momentum.pdf.

Koizumi, Shinjiro. 2020b. "Petersberg climate dialogue XI 2020." https://www.bmu.de/en/petersberg-climate-dialogue-xi/.

Korea News Gazette. 2020. "FM says COVID-19 recovery should be aligned with push to tackle climate change." *Korea News Gazette*, 24 June 2020. https://login.ubproxy.ub.uni-heidelberg.de/login?qurl=https://advance.lexis.com%2fapi%2fdocument%2fcollection%2fnews%2fid%2f6070-H5N1-JCH9-G15T-00000-00%3fcite%3dFM%2520says%2520COVID-19%2520recovery%2520should%2520be%2520aligned%2520with%2520push%2520to%2520tackle%2520climate%2520change%26context%3d1516831.

Korea, Republic of. 2020a. "Keynote address by President Moon Jae-in at Presentation of Korean New Deal Initiative." 14 July 2020. https://www.korea.net/Government/Briefing-Room/Presidential-Speeches/view?articleId=187807&categoryId=111&language=A020101&pageIndex=7.

Korea, Republic of. 2020b. "Opening remarks by President Moon Jae-in at 6th Emergency Economic Council Meeting." 1 June 2020. https://www.korea.net/Government/Briefing-Room/Presidential-Speeches/view?articleId=186051&categoryId=111&language=A020101&articleId=null&pageIndex=9.

Korea, Republic of. 2020c. "Opening remarks by President Moon Jae-in at ROK-EU Summit." 30 June 2020. https://www.korea.net/Government/Briefing-Room/Presidential-Speeches/view?articleId=187162&categoryId=111&language=A020101&articleId=null&pageIndex=8.

Korea, Republic of. 2020d. "Remarks by President Moon Jae-in while visiting Korean New Deal and Green Energy Site." 17 July 2020. https://www.korea.net/

Government/Briefing-Room/Presidential-Speeches/view?articleId=187857&categor
   yId=111&language=A020101&pageIndex=7.

McQueen, Cormack. 2020. "A green deal to remould the shape of our politics; ,
   'Nearly there': Programme to be signed off by three party leaders this morning."
   The Irish Independent, 15 June 2020. https://advance-lexis-com.ubproxy.ub.uni-heid
   elberg.de/api/document?collection=news&id=urn:contentItem:604N-7B31-DYTY-
   C2KH-00000-00&context=1516831.

Ministry of the Environment, Japan. 2020a. "Japan's Roadmap to 'Beyond-Zero'
   Carbon." Japan: METI. https://www.meti.go.jp/english/policy/energy_environm
   ent/global_warming/roadmap/index.html.

Ministry of the Environment, Japan. 2020b. "Ministerial meeting of the 'online
   platform' on a sustainable and resilient recovery from COVID-19 and the
   website 'Platform for Redesign 2020'." Press release 25 August 2020. Ministry of
   the Environment, Government of Japan.

Ministry of the Environment, Japan. 2020c. "Online platform on sustainable and
   resilient recovery from COVID-19." https://platform2020redesign.org/.

O'Sullivan, Kevin. 2020. "Former president calls for stronger response to climate
   change crisis." *The Irish Times*, 7 March 2020. https://advance-lexis-com.ubproxy.
   ub.uni-heidelberg.de/api/document?collection=news&id=urn:contentItem:5YC9-S
   WG1-JC8Y-800M-00000-00&context=1516831.

Prime Minister, UK. 2020. "'Build build build': Prime Minister announces new deal
   for Britain." Press release. Published 30 June 2020. https://www.gov.uk/governm
   ent/news/build-build-build-prime-minister-announces-new-deal-for-britain.

Sunday Observer. 2020. "Protecting environment and human health amid Covid-19 fears."
   Sunday Observer (Sri Lanka), 14 June 2020. https://advance-lexis-com.ubproxy.ub.
   uni-heidelberg.de/document?crid=dde08d00-3840-48f4-9957-3c0e0c3f9610&pddoc
   fullpath=%2Fshared%2Fdocument%2Fnews%2Furn%3AcontentItem%3A604G-
   7FK1-JDKC-R4NC-00000-00&pdcontentcomponentid=338356&pdmfid=1516831
   &pdisurlapi=true.

UK Department for Business, Energy & Industrial Strategy. 2020a. "Resilience in
   light of COVID: Climate action on the road to COP26." Keynote address by
   COP26 President, Alok Sharma. Published 3 July 2020. https://www.gov.uk/gov
   ernment/speeches/resilience-in-light-of-covid-climate-action-on-the-road-to-cop26.

UK Department for Business, Energy & Industrial Strategy. 2020b. "The sustainable
   recovery, investor collaboration on COVID-19 recovery and the climate emergency."
   Keynote address by Alok Sharma, Business Secretary and COP26 President at a
   London Stock Exchange Webinar. Published 17 June 2020. https://www.gov.uk/
   government/speeches/the-sustainable-recovery-investor-collaboration-on-covid-19-r
   ecovery-and-the-climate-emergency.

UK Environment Agency. 2020. "Business as usual won't tackle the challenges we
   face, warns Environment Agency." Press release, 9 July 2020. https://www.gov.uk/
   government/news/business-as-usual-won-t-tackle-the-challenges-we-face-warns-enviro
   nment-agency.

Vetter, David. 2020. "South Korea embraces EU-style green deal For COVID-19
   recovery." *Forbes*, 16 April 2020. https://www.forbes.com/sites/davidrvetter/2020/04/
   16/south-korea-embraces-eu-style-green-deal-for-covid-19-recovery/#283e00765611.

World Economic Forum. 2020a. "Billions for sustainable investments – Germany's
   plan for a green recovery." Accessed 24 October 2020. https://www.weforum.
   org/agenda/2020/07/germany-green-recovery-billions-sustainable-investments/.

World Economic Forum. 2020b. "COVID-19 amplifies inequality. Fight back with long-term thinking." Accessed 24 October 2020. https://www.weforum.org/agenda/2020/07/covid-19-amplifies-inequality-fight-back-with-long-term-thinking/.

World Economic Forum. 2020c. "World vs virus podcast: An economist explains what COVID-19 has done to the global economy." Accessed 25 September 2020. https://www.weforum.org/agenda/2020/09/an-economist-explains-what-covid-19-has-done-to-the-global-economy/.

World Health Organization, United Nations (WHO). 2020a. "Timeline: WHO's COVID-19 response." Accessed 23 October 2020. https://www.who.int/emergencies/diseases/novel-coronavirus-2019/interactive-timeline.

World Health Organization, United Nations (WHO). 2020b. "WHO Coronavirus Disease (COVID-19) Dashboard." Accessed 23 October 2020. https://covid19.who.int/.

16 Insights for policymakers

In addition to its academic contribution, this study has provided numerous insights on how health adaptation policies in different countries have developed over time, which best practices exist, and which pitfalls should be avoided. Moreover, the study shows which role international organizations play for countries when it comes to health adaptation to climate change and how projects should be designed to make them the most effective and sustainable at the same time. Last, but not least, the book has extensively elaborated on how complex international policy challenges with high levels of uncertainties can be tackled, how inequalities between countries due to

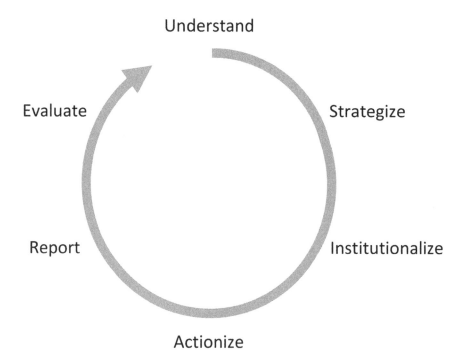

Figure 16.1 The 6-Step Policy Development Circle

varying levels of socioeconomic and institutional capacity can be balanced out through international cooperation, and which role epistemic communities can play to support decision making in this field. Consequently, the analysis has led to a number of specific and actionable policy recommendations, which will in the following be briefly presented.

Understanding

To better understand "wicked problems," such as health adaptation to climate change, special expertise and experience is needed. This includes a general understanding of how climate change affects specific health determinants as well as exact/precise vulnerability assessments that analyze which parts of the population are affected how and why. Importantly, the assessments should not only focus on very direct climate change related health risks, such as those related to extreme weather events, but also take into account the interdependencies between local, regional, and global developments as well as the nexus between climate change and other key areas such as migration or conflicts. Moreover, it is important for decision makers to have an overview of which risks can be expected when, and with what intensity, so that they can prioritize which response is needed immediately and in the mediate and long-term future.

Since climate change and health is a rapidly developing and still relatively novel policy area, many states lack the necessary research capacity and institutions to fully understand the risks climate change poses to the health of their population. As a consequence, it is crucial for them to draw on the expertise of the epistemic and international community, such as universities, think tanks, advisory agencies, and international organizations, while further developing their internal capacity. In addition to in-person expertise, they can benefit from digital solutions, such as knowledge-exchange platforms, virtual conferences, and exposure assessment tools, including those developed by the World Bank.

Strategizing

Once a thorough understanding of climate change related health risks in the specific regions has been established through first conferences, research projects, and vulnerability assessments, it is instrumental to develop sound, sustainable, and actionable strategies and plans. This does not mean that every single step on the path towards sustainable health adaptation is planned out from the beginning. However, the major challenges and corresponding solutions, instruments, and best practices should be identified, clear recommendations and roadmaps should be developed, and scenarios for different potential future pathways should be constructed in order to increase the preparedness and resilience of the affected areas and populations.

The effectiveness of and public support for the respective adaptation strategies can be largely increased through the inclusion of various stakeholders in the strategy development process through co-creation techniques. If representatives from civil society, the public and private sector, and academia work on such strategies together, a large variety of perspectives can be included, and more risks and co-benefits can be anticipated than if strategies are developed by more homogenous groups. Such transdisciplinary initiatives require outstanding moderation techniques and motivated stakeholders, but they can make all remaining steps on the road towards sustainable health adaptation much faster and easier since opposition against the planned measures is likely to be much lower if all relevant stakeholders are involved from the outset.

Institutionalizing

Building institutions and capacities to tackle "wicked problems" is the most essential step on the path towards sustainable health adaptation measures. Before delving into the details of how institutions can be built, however, it is important to understand that institutions are not merely organizations like the United Nations or research institutes, but rather represent both formal and informal rule-based systems that interact with each other, define certain behavior principles, and create expectations (Keohane and Nye 1989). With regards to health adaptation to climate change within states, institutions can be ministerial departments and agencies but also laws, contracts, and much more. Best practices in terms of the institutionalization of health adaptation to climate change include the UK's Climate Change Act from 2008, which led to the creation of new expert positions on climate change and health within governmental agencies and ministries. Another primary example constitutes the establishment of the Korea Adaptation Center for Climate Change (KACCC) in 2009, which integrates and coordinates the RoK's adaptation work.

Since the nexus between climate change and health constitutes a highly complex long-term policy challenge, and political agendas are often dominated by crises, institutions become necessary to ensure that governments do not lose track of such challenges and create the capacity to understand such risks and take effective actions. To ensure the adoption of balanced, effective, and up-to-date adaptation measures, it is necessary to design institutions and processes in a sustainable manner, which entails a clear mandate, a coherent integration into existing frameworks and actions, as well as independence of project-based funding. When international organizations, NGOs, or other states provide funding for health adaptation projects, it is important to design them with a long-term perspective in mind. This includes a focus on capacity building, knowledge and technology transfer, and training of the next generation of experts within the respective country.

Actionizing

For health adaptation to climate change to be effective, it is of utmost importance to follow recommendations, strategies, and plans with concrete adaptation-level measures. These can include policies, capacity building measures, and educational activities as well as very tangible actions, such as the building of new infrastructure, the introduction of new behavioral practices, the installment of monitoring and warning systems, or the provision of financing instruments. For adaptation to work and receive support by voters and interest groups, however, it is pivotal to ultimately move beyond recognition-land groundwork-level initiatives, and implement concrete adaptation-level measures, such as enhanced infrastructure, warning systems, or new practices.

Considering that not all states have the same capacities to take effective action against climate change related health risks, they can often benefit from international support, either through international organizations and NGOs or directly from other states. Such support can help to reduce inequalities and prevent global challenges from accelerating even further. Nonetheless, for international support mechanisms to be successful and sustainable, they have to be designed in close cooperation with experts from the respective countries and with their goals and perspectives in mind to ensure that project ownership can be transferred to policy ownership once external actors are no longer active in the respective country.

Reporting

Reporting on national adaptation measures is an essential part of international climate change reporting frameworks, such as National Communications to the UNFCCC. Moreover, almost every country in the world nowadays has national adaptation plans that follow similar structures. Regularly publishing national adaptation strategies and plans has thus become an international standard and states with more innovative, comprehensive, and effective strategies than others can gain a leading role in the international community and thus international reputation. In addition to the international component, national-level reporting is of high importance for internal politics since it can set the standards for regional and local adaptation strategies and build the foundation for an ongoing dialogue with citizens and representatives from various interest groups on how to best adapt to "wicked problems," such as the climate change and health nexus.

Evaluating

Due to the fact that the intersection between climate change and health is a rapidly moving policy field, evaluating and updating current measures based on new academic findings, changes in the international community, new national priorities, and emerging risks is essential to effectively protect populations from climate change related health risks. When doing so, national experts can rely on publications and other sources provided by international organizations and epistemic communities. At the same time, external and independent experts, such as researchers and policy advisors, should be integrated into the evaluation process to ensure unbiased and transparent evaluation procedures and identify pathways for enhancement.

For more detailed and country-based recommendations, including risk analyses and strategy development consultancy, feel free to contact me at jungmann@momentumnovum.com.

References

Keohane, Robert O., and Joseph S. Nye. 1989. *Power and Interdependence*, 15th ed. Boston/ Glenview/London: Scott Foresman.

Index

Printed in the United States
by Baker & Taylor Publisher Services